Chapter 1

Introduction

1.1   General Information

There are many books in the library which deal with the subject of alternate energy. Some are useful, but many treat the subject in a wishful way ... alternate energy is regarded as almost free energy. With this or that homebuilt apparatus, easy energy is ours for the taking. While it makes for romantic reading, the practical results of such an attitude leave much to be desired.

Our first task therefore, is to set the facts straight. You cannot hope to produce energy as cheaply as a public utility which has the economics of mass production in its favor. If you wish to generate sufficient energy for your needs, expect that energy to cost more than buying it from the utility company. We don't mean to imply that you can't actually lower your energy costs with an alternate energy system, because you may do so. The cost reduction, however, will come about from lowering energy consumption, not from paying less for the energy. Conservation is the mandate of an alternate energy system, and a lesson we all need to learn.

Alternate energy systems cost dearly to install ... despite the appeal of free energy from wind, waves or the sun, tapping that energy involves costs. Equipment to harness such energy sources must be finely tuned. Crudely built homemade devices are unlikely to yield the efficiencies necessary. Even electrical generation equipment that is built to good engineering practices must face the limits of physical science. There is no free energy.

We use the term *alternate energy system* to mean any system other than the public utilities. This is not wholly consistent with other's use of the same term to designate energy systems which rely solely on the sun or wind. An alternate energy system in this book is one that includes sun and wind sources, but may also include DC alternators and/or AC generators which run from small engines. Refrigeration is also part of the alternate energy system. By the same token, energy for heat and hot water are part of an alternate energy system, but we don't include information on those subjects.

1.2   Why Alternate Energy?

We can think of several reasons why you should consider an alternate energy system. They are freedom, freedom and freedom. You may be living in a boat, at a place of your choosing. There are no power lines to tap into ... what you have on board is all you have. You may be residing in a land yacht at the end of a scenic road. Your home may be a *mountain mansion* of one room or more, far from the nearest utility. We think the major appeal of an alternate energy system lies in the independence which it provides. Such a system is not free, but is instead the price of freedom, and in a world of city throngs, a small price to pay.

1.3   Making it Work

Virtually any level of alternate energy can be made to work. You may simply elect to replace the utility company with your own *AC* generator which runs 24 hours a day. Or, you may elect to use a small solar panel which provides power to your only electrical appliance ... a shortwave radio. This book is not written for those extremes. Instead, we focus on a system that provides essential lighting and refrigeration energy, but lacks the gross amount of energy that is typically wasted by inert consumers day in and day out.

Most of the alternate energy systems currently in place were not designed. The builders of them have simply collected an assortment of equipment which they interconnected. The results are what can be expected from lack of engineering ... poor performance. By nature, however, alternate energy systems are called upon to serve a diverse set of needs. Each system cannot be specifically engineered. If the user expects to achieve good performance, then he or she must learn the essential facts which must fit together without mismatch. Making it work well is quite dependent on a knowledgable user.

Even systems constructed by boat and RV manufacturers are not designed, but are a collection of outdated equipment and vague ideas about what level of power is necessary. It is all too common to see hundreds of Amp-hours of electrical loads in a system with a puny alternator charging $50 worth of batteries. Thousands of boats are equipped with small *DC* refrigerators that the owner turns off on the second day away from the dock because of dead batteries. This same owner will have the latest in exotic sail material, foul weather gear, self-tailing winches, and a plethora of wonderful electronic gadgets to make the trip a safe one. Who in their right mind could think that a $100 energy system is adequate in this scene?

Has this strange set of circumstances developed because the energy system is usually hidden from view and not subject to Madison Avenue sales pitches? Are the manufacturers only interested in enough energy to get the engine started one time ... when the new owner takes delivery? Are we users just too familiar with the electrical systems in automobiles, and can't imagine that they aren't sufficient for demands of a *liveaboard* situation?

The answer to these questions is naturally a complex answer involving many aspects not even considered here. But you don't need to answer the question in any other way than you already have. By deciding to read this book, you have decided to find out for yourself what it takes to live on 12 Volts. In doing so, you have taken the first step to changing the energy systems that we now see in boats, *RV*s, and remote homes.

1.4   How to Use This Book

Some readers may find this book easy reading. Others may have to struggle to glean the information needed for their circumstances. On the one hand, we have tried to be as informative as possible without sinking in a mire of technical data. On the other hand, we have attempted to be complete, providing all the facts necessary for a thorough understanding of the subject.

Given these conflicting goals, we recommend that you don't read this book until you have read it several times. This is a pun, of course, but it serves to illustrate our real recommendation. Read the book here and there until you have a general idea of the subjects discussed. Attempt to fit the information in the book to your own needs and desires for an energy system. Finally, when you have developed a structure for the kind of system you think is appropriate, study those parts of the book which deal with those components of interest.

1.5   About the Contents

As the front and back covers indicate, this is a book about electrical and refrigeration energy systems. An effective alternate energy system must have a storage of energy. Stored energy needs a replenishing source. Limited energy resources must be consumed by intelligent and dedicated users.

In a nutshell then, this book is about energy storage, energy sources, and the wise use of energy. When these elements are matched and assembled, we call it the Balanced Energy System. Such a system is made up of many components, each of which must be understood before application. In general, each component is the subject of a chapter. Given below is a short synopsis of the various chapters.

1.5.1   Chapter 2

Our system makes use of lead acid batteries to store energy captured during its available periods. Such energy may be derived from sun, wind, water, *DC* alternators, or *AC* operated battery chargers. Since these energy *Sources* provide power on an intermittent basis, some storage is necessary. Among types of batteries, the lead acid battery provides the greatest amount of storage per dollar of investment.

The lead acid battery tolerates all manner of abuse while still providing a measure of service. Because the battery doesn't immediately quit when abused, but rather loses some of it storage capacity, the effects of abuse are not readily observed. When the user discovers that the battery does not provide enough storage, more batteries are added to the system. Eventually a user, determined to use batteries, will have many more of them in the system than necessary ... all of them operating at a fraction of their potential.

In Chapter 2, the lead acid battery is discussed in depth. Only when the lead acid battery is understood can it be applied intelligently. A high quality deep cycle battery will provide many years of excellent service if treated with reasonable care. Because the storage of electrical energy is so important to the success of the system, we have given batteries extensive coverage.

Before reading Chapter 2, it may help to consider the following questions.

- What happens inside a battery when it discharges and charges?

- How much energy can be stored?

- How is storage capacity measured?

- How can capacity remaining be determined?

- When is the best time to recharge?

- What charging techniques yield optimum performance?

- How fast can a battery be charged?

- How should a battery be discharged?

- What kind of regular maintenace is required?

- What pitfalls must be avoided?

1.5.2   Chapter 3

In Chapter 3 we discuss a new breed of lead acid battery, the immobilized electrolyte, pressure sealed units. Their sealed construction makes them suitable for use inside living quarters of the home. They are truly maintenance free, are position independent, absorb a faster charge and have other benefits. Though still more costly than the conventional lead acid battery, builders of high quality energy systems will elect to use them.

1.5.3   Chapter 4

Chapter 4 is devoted to the *DC* alternator. Whether the alternator is driven by a small engine, the wind, or water, it is a very cost-effective source of energy. For charging batteries fast, the *DC* alternator is much more efficient than an *AC* generator with a battery charger. It seems that everyone makes small *AC* generators for portable power. *AC* generators may be suited to the occasional back country tourist, but the *DC* alternator is more in tune with the stored energy system.

While the *DC* alternator itself is quite appropriate, the standard regulator attached to the alternator has no place in a high performance energy system. The fact that it doesn't charge batteries properly will become apparent from Chapter 2.

1.5.4   Chapter 5

The *AC* operated battery charger is covered in some detail in Chapter 5. Despite the various circuit topologies used in existing battery chargers, none is really suited to charge and maintain deep cycle batteries. The unregulated variety must be disconnected from the battery when a full charge is reached. Though we advocate active awareness of your electrical system, monitoring a battery charger is more than can be expected.

Regulated battery chargers are constructed with two basic regulation methods. The cheapest, and most reliable means is called ferroresonant regulation. This type of regulation is grossly inefficient and ineffective for deep discharge replenishment. User ignorance coupled with attractive packaging by some manufacturers of ferro-resonant chargers keep sales high despite the unsuitability of the product.

Regulation is also achieved with active electronic circuits which sense battery voltage and adjust charger output. Alas, in the eyes of marketeers and designers everywhere, users fall in one category ...stupid. These idiots must be protected from their stupidity and the way to do that is to fix the output of the battery charger at a constant voltage. Never mind that such a voltage setting is always a serious compromise, leading to poor charging performance and premature battery failure. It's idiot proof.

1.5.5 Chapter 6

Solar panels are the subject of Chapter 6. Solar panels used in sufficient quantity can supply all your energy needs. Only cost stands in the way. When we purchased an 1800 Watt generator some years ago, the equivalent solar panel array would have cost about $20,000. Today's costs are not much less. But it is unfair to compare only the original costs, and the fact remains that with a properly designed system and an aware user, much of the required energy can be obtained from the sun.

1.5.6 Chapter 7

Wind and water, the subject of Chapter 7, have been used by humanity for many centuries. Modern humans make good use of the energy available in falling water, and in some locales, have started to harness the wind. It only stands to reason that an individual energy system can benefit from wind and water. The costs are high, however, on a Watt per dollar basis. Wind in particluar is usually too variable to provide a reliable energy source. Water has to fall a long way to yield much energy, but a moving boat can take advantage of a water driven generator.

1.5.7   Chapter 8

In Chapter 8 we discuss the place of AC in the alternate energy system. How to use it safely and effectively is covered. The AC motor can be manufactured less expensively than a comparable DC motor. It was this fact, and the fact that AC can be transmitted long distances over wires that led to universal acceptance of AC power. Most of the appliances we use are AC operated. While an alternate energy system does not require AC power, it is nice to have occasionally.

There is more to using a small AC generator than starting it and plugging in. When used to charge batteries, the type of charger will affect the charging performance. Used properly, a small AC generator compliments the DC system.

DC to AC inverters are discussed, particularly their operating waveforms and power efficiencies. Typical AC loads are given.

1.5.8   Chapter 9

Most of us regard refrigeration as a basic neccessity. Chapter 9 deals with the principles of refrigeration systems. Good refrigeration habits are presented. The role of insulation is stressed, and the means of calculating the effectiveness of insulation is given.

It is likely that refrigeration will be the major energy consumer in your system. As such, the choice of a refrigeration method is critical. Small *cold* leaks in the refrigeration system, whether by faulty design or ignorant usage, can spell doom for the rest of your energy system.

1.5.9   Chapter 10

In Chapter 10 we discuss the details of compressor refrigeration systems. Included are engine driven systems, and DC operated systems. The continuous cycle versus intermittent cycle is explained. Important aspects of holdover plates and eutectic brines are presented. Compressors, condensers, evaporators expansion valves and other components are illustrated.

1.5.10   Chapter 11

Chapter 11 deals with non-compressor refrigeration including thermo-electric modules and absorption systems.

1.5.11   Chapter 12

Chapter 12 is by far the most difficult of the chapters. This chapter provides the *design rules* that must be followed to design a compressor refrigeration system. Subjects include:

- sizing refrigeration load
- sizing an evaporator
- sizing a condenser
- sizing a compressor
- sizing an expansion valve
- sizing a holdover plate
- calculating refrigerant velocity
- calculating the time to freeze a eutectic brine
- calculating cooling water requirements

1.5.12   Chapter 13

The subject of Chapter 13 is the Balanced Energy System. This is the chapter that brings together the prior chapters into a single system. A refrigeration system is described, and the load that it presents to the electrical system is determined. Other electrical loads are described, and a total daily power requirement is determined. Batteries to store the needed energy are selected. Alternators, solar panels, wind generators, and battery chargers to replenish the consumed energy are described.

To be energy balanced, Batteries, Sources and Loads must fit together in *golden ratios*. By selecting the energy system components based on optimum performance ratios, the adequacy of the system is assured. With the golden ratios of Chapter 13, you, the user, can assemble an energy system with the reasonable expectation that it perform up to your needs.

1.5.13   Chapter 14

Chapter 14 is a vital chapter for boaters only ... the subject of *AC* safety and electrolysis. The mechanism of electrolysis is described, and the results of connection to the *AC* safety ground is presented. Risks of electrocution by ungrounded *AC* are tabulated, and the means to minimize shock hazard and electrolysis is given.

1.5.14   Chapter 15

This chapter is a potpourri of subjects ranging from instrument display types, resolution and accuracy of measurements, Charge/Access Diodes, heatsink surface area requirements, noise filters and proper power and ground connections. Also included is a short discussion about the use of isolation diodes and their impact on the charging voltages of two batteries.

1.5.15   Chapter 16

Chapter 16 is a reference to electrical and electronic circuits and symbols. Units of measure are discussed, and historical information regarding the naming of these units is presented.

This was the most interesting chapter to write. Despite a lifetime spent in close proximity to the latest technology, with a new subject to learn on a daily basis, we can only stand in awe of the people who first began to assemble the basic scientific data that is the cornerstone of modern humanity.

Chapter 2

Liquid Electrolyte Batteries

2.1 Historical Information

In the year 1790, Luigi Galvani discovered that electrical current would flow between two dissimilar metals placed in a conductive liquid. At the time, he was probing a frog with a copper probe, and noticed that the frog's muscles reacted convulsively. The frog was hung from an iron hook, and there was electrical current between the copper and iron. Galvani didn't understand his discovery but the reaction of two metals in an electrolyte has become know as galvanic action.

Ten years later, Alessandro Volta correctly deduced that the frog's muscle twitches, observed by Galvani, were the interaction of copper and iron, acted upon by electrolytic action of the frog's blood. Battery chemistry was on its way, first as an art, and later as a scientific technology.

In the early 1830s, the great scientest Michael Faraday performed the quantitative analysis of electrochemical reactions in a battery. In the process, he developed the terms *electrolyte, electrode, anode* and *cathode.*

The first practical battery was developed in 1836, but it could not be recharged. The rechargeable battery has been with us since 1859, but at that time the only way to recharge the battery was by discharging non-rechargeable batteries ... the generator hadn't been perfected yet. Later, around 1880, Thomas Edison was selling elec-

11

trical energy by day, and recharging his batteries with a generator at night when power demand was lowest. Today, considerable money is spent on research for better means of using batteries as *load leveling* storage for large power plants, as first practiced by Edison.

In 1895, the land speed record of 68.5 *MPH* was held by an electric car, powered by a lead acid battery. Charles Kettering[1] invented the electric starter in 1911, and installed it in the 1912 Cadillac, displacing the hand crank which was then used to start automobiles. Since that time car batteries have become civilization's biggest consumer of lead.

2.2   General Information

The conventional lead acid batteries of today are made essentially the same way they were in 1911 ... two dissimilar metals are placed in a conductive liquid. Lead dioxide plates, with intervening lead plates are submerged in sulphuric acid to form a cell. A single lead acid cell produces about 2 Volts. A cell may have as many as a dozen pairs of plates, though most don't have this many. A 12–Volt battery is made by connecting six, 2–Volt cells in series.

To prevent short circuits within the cell, the individual plates are separated from each other. The quality of the separator is very important. It prevents short circuits by inhibiting migration of active material between the positive and negative plates, but allows a free flow of electrolyte. Quality separators are made of fiberglass or microporous rubber as opposed to paper. Battery construction details vary according to the quality of construction and the life expectancy of the battery. Deep cycle batteries have thicker plates and are better supported with thicker grids and separators than those of the automotive battery. Whereas an automotive battery may only withstand several complete discharges, the deep cycle battery will tolerate several hundred.

About 25 years ago, a sealed *thixotropic gel* lead acid battery was developed. We make note here the difference between the sealed batteries now used in automobiles, and the thixotropic gel type of sealed

---

[1]Kettering founded Dayton Engineering Laboratories Company, Delco, which merged with other companies to form General Motors.

batteries. The sealed automotive batteries are nothing more than conventional lead acid technology, partially sealed with enough liquid electrolyte to last for the short life of the other materials. The thixotropic gel batteries do not have liquid electrolyte, rather the acid is captivated in a gel. (Other similar sealed batteries captivate the acid inside the fiberglass separators).

At last, the sealed batteries have become economically feasible for alternate energy systems such as those on boats. There are enough significant differences in operating characteristics, and performance of this *newcomer* technology to warrant a separate chapter to describe them. The remainder of this chapter, therefore applies to conventional liquid electrolyte batteries. The sealed lead acid battery is much like its predecessor, so this chapter should be read for historical reasons even if you have already chosen immobilized electrolyte batteries.

2.3 Electrochemical Reactions

A discharging battery is converting lead dioxide to lead sulfate, and sulphuric acid to water. This process is reversed during charging. It is important to remember that while the electricity produced by chemical reactions follow electrical laws at the speed of light, the reactions inside the battery are chemical reactions that take time. In particular, diffusion of the acid throughout the battery is relatively slow, so heavy load currents deplete the acid in the immediate vicinity of the plates. A short rest allows acid diffusion to take place and this accounts for the *recovery* phenomenom often experienced with lead acid batteries. The recovery experience is common to many of us ...our old car won't start and runs down the battery. After a break to cool our temper, the battery is recovered enough to start the car. We can take advantage of the recovery effect by discharging batteries intermittently, rather than continuously.

Electrolyte diffusion during charge presents a practical limitation as to how fast a battery will accept a charge. To grasp the significance of diffusion, consider the case of a tea bag in a cup of water. If the water is not agitated, how long does it take for the tea to diffuse throughout the water? There is some agitation of the acid in the battery, but not sufficient to bring about rapid diffusion or mixing of the acid. As you might expect, thick plate deep cycle batteries are slow to recover from fast discharges because of longer acid diffusion times through the thick plates. Thicker plates must also be charged at a slower rate.

2.4   Specific Gravity

The work of the battery is done by three elements, the positive plate the negative plate and the electrolyte. The positive plate is made of lead dioxide, while the negative plate is lead. Electrolyte is a mixture of sulphuric acid and water, and its reaction between the positive and negative plate produces electrical energy.

The concentration of acid to water in the electrolyte is known as *specific gravity*. In a typical deep cycle battery, electrolyte weighs 1.265 times as much as water. The decimal point is dropped and the specific gravity, *(SG)* becomes 1265 *points*. A heavier (say 1.300) electrolyte can provide more capacity from a battery, but at a reduced lifetime.

Specific gravity for fully charged deep cycle batteries falls in the range of 1250–1280. The same batteries flat discharged will register 1140–1170.

Specific gravity is measured with a hydrometer. It is a weighted float which is calibrated with a readable scale that indicates how deep the float sinks into fluids. The scale reads *SG* points directly. *SG* measurements are stated at $77°F (25°C)$ and vary with temperature. At lower temperatures, the resistivity of electrolyte is higher, thus a cold battery exhibits reduced capacity. When taking *SG* measurements temperature compensation should always be applied.

Healthy batteries will have uniform specific gravity measurements (taken with a hydrometer) from one cell to the next. A difference of 30 points indicates that the battery could use an equalization charge. If you get readings from cell to cell that vary by 50 points or more, buy a new battery.

2.5   Battery Capacity

How much energy a battery can store is known as its capacity. Capacity is stated in units of Ampere-hours (Ah). Often, we tend to think of capacity in terms of fuel tank capacity, but this concept can be misleading. While you always get ten gallons from a ten gallon tank, battery capacity varies according to the rate of discharge. That is, the faster you remove energy from a battery, the less of it you can get. For example, a battery that yields 5 amperes of current for a period of 20 hours can be specified as a 100 Ampere-hour battery. This same battery can only provide 50 amperes of current for one hour, not for two hours as you might expect. Figure 2.1 illustrates the effect of discharge rate on battery capacity. For a one hour discharge, there is about 45% as much energy as the 20 hour rate. As shown, 100% capacity is achieved at a 20 hour discharge rate. A battery discharged in 1 hour will only yield about 45% of the capacity at the 20 hour rate. Note how rapidly the capacity falls when the discharge rate is high ... thick plates mean slow diffusion of electrolyte.

Figure 2.2 shows the effect of temperature and discharge rate on available capacity. This data is applicable to high quality deep cycle batteries with thinner plates than those shown in Figure 2.1. The curves are for 20 hour rate (0.05 of capacity), 10 hours, 4 hours and 1 hour. Note that the battery holds up to higher loads better than the thick plate unit shown in Figure 2.1. Cycle life on the thinner plate battery will not be as high as the thicker plate unit.

Specifying battery capacity at the 20 hour rate is not standard. Some manufacturers specify capacity at a 100 hour discharge rate. Other batteries are specified at 5, 6 and 8 hour rates. We are more interested in the longer term capacity ... from 20–100 hours.

Battery capacity is also a function of temperature. The lower the temperature, the less the capacity. As shown in Figure 2.2, you can plan on 3–5% loss for every $10°F (5.5°C)$. Capacity is specified in the U.S. at $77°F (25°C)$.

While capacity does increase with temperature, batteries should never be operated (charged or discharged) above $120°F (50°C)$. Batteries that are to be operated continuously at high temperatures should have the acidity of the electrolyte reduced somewhat to ex-

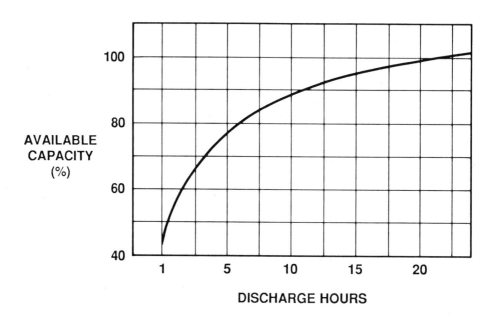

Figure 2.1: Capacity versus Discharge Rate

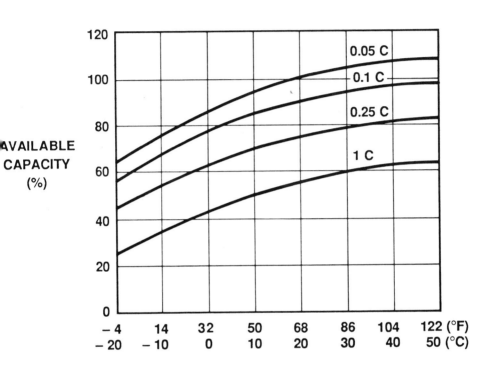

Figure 2.2: Temperature and Discharge Rate versus Capacity

tend their operating life, although their capacity will be reduced by about 10%.

Capacity actually increases in the early stages of using a new battery ... as much as 20% if charged properly and not discharged rapidly on a regular basis. Most manufacturers rely on this increased capacity to meet their claims, thus a new battery may have 10–20% less capacity than advertised.

As the battery ages, capacity falls, primarily due to sulfate clumps which form on the lead dioxide plates during discharge and don't get broken up on re-charge. Later we'll explain how you can extend the life of your batteries by *equalizing* them.

When the capacity of a battery has fallen to about 75–80% of its new capacity, you should replace it. It starts losing capacity fairly rapidly at that point, as ever larger sulfate crystals start to surround and isolate remaining active material.

The capacity of a 12–Volt battery can be measured by performing a discharge with a known current and timing the length of time it takes to discharge to 10.5 Volts. This test should be performed with a discharge current that is close to the average current that the battery will support for 20 hours. More details about capacity testing are presented later.

As mentioned, capacity is usually stated in terms of Ampere-hours, with 20 hours being the normal discharge period. There is a theoretical **capacity-to-weight** ratio limit. In practice, the capacity is far below the limit. You can expect to get about 0.96–1.35 *Ah* per pound of 12–Volt battery. We realize that these numbers are not consistent with the capacity claims made by some manufacturers, but the very limited capacity testing which we have performed, while far from conclusive, supports the published data about capacity-to-weight ratio. The wide range of *Ah* per pound is due not only to slight differences in technology and manufacturing practices, but can often be attributed to the case material that houses the battery. Older designs still use phenolic (hard, brittle and heavy) cases, while newer designs (especially sealed units) use lightweight and extremely durable polypropolene cases. Lightweight cases can mean more active material in the battery, for any given size.

Before buying batteries, take into account the capacity-to-weight

ratio presented above. You probably should avoid those hard brittle
cases for safety reasons. Be concerned about the lightweight batteries
that appear to have a high capacity rating. To know how much actual
capacity a battery has, you will have to perform a capacity test.

You should not be lead astray by ratings of cold cranking amps,
(CCA), or *reserve* capacity. These are ratings applied to engine starting batteries, not deep cycle units.

Determining[2] the capacity you need is a matter of tabulating the
current draw of all the electrical equipment on board, and then multiplying each current by the number of hours of daily use. Now add
the (Amps times hours) figures for the total Amp-hours per day of
consumption. How many days between charging do you want? For
each day desired, multiply by the daily consumption. The total you
get is the *Ah* rating of *one* battery. We say one battery, because we
assume that you are not going to discharge more than 50% of rated
Ah. More details about calculating required capacity is given later in
the book.

In general, you want as much capacity as you can afford, and can
carry, but boat owners pay for much more capacity than they actually
use due to poor charging methods. We know boats with 1000 *Ah* of
batteries aboard that may still be good for 300 *Ah*. Since the batteries
were never fully charged, permanent sulfation has taken its toll. Extra
ballast is cheaper when using lead ingots.

You should perform a capacity test on batteries 3 years old or
more. (Testing your brand new batteries is not a bad idea either.
Why rely on a manufacturer's claim that may not be accurate.) A
capacity test starts by charging and equalizing the batteries first.

Battery capacity can be determined by applying a load current
that is approximately 5% of the declared capacity. Load current must
be measured with an accurate digital meter via a precision low voltage
shunt in series with the battery. With a 5% load applied to a fully
charged battery, the voltage will fall to 10.5 Volts in 20 hours ... that
is if the declared capacity is a valid specification. Since you can't
count on that, you will need to periodically observe the voltage on

---

[2]For those that have access to a PC, or other compatible computer, we offer
a program which calculates the daily *Ah* required, and appropriate size charging
equipment for your system.

Discharge Hours	Capacity (percent of rating)
30	105
20	100
10	89
5	78

Table 2.1: Discharge Rate vs. Capacity

the battery. Whenever it falls to 10.5 Volts, turn off the load and multiply the load current times the number of hours that the load was connected. The result is the capacity at the measured test current. If the test didn't go for 20 hours, you need to adjust the number to determine 20 hour capacity. Use Table 2.1 as explained below.

Suppose you tested a battery rated at 100 Ah. Your test current would be 5 Amps. If the battery only produced for 10 hours, you got 50 Ah at the 10 hour rate. Dividing this by 0.89 from Table 2.1, the 20 hour rate becomes 56 Ah. Refer also to Figure 2.2.

Besides a periodic capacity test, you should also test the condition of the batteries. This is easily done on a fully charged battery. Set the voltage to 14.4 Volts and monitor the battery current. It should be 1% or less of the Ah rating. Remember, the battery must be fully charged before the current will fall below 1% of its Ah rating. If it doesn't reach that level after 16–20 hours, the battery is starting to show signs of age.

2.6   Percent Capacity Remaining

You will get more power for your dollars if you never discharge your batteries below 50% of their capacity. Discharges to less than 20% of capacity can be very damaging to a battery, but this is common practice, and, as a result, new batteries are purchased regularly. Energy cost versus depth of discharge is illustrated by Figure 2.3. The factors affecting cost are the price of the batteries, as well as the costs associated with charging them. Note that cost is least around a 50%

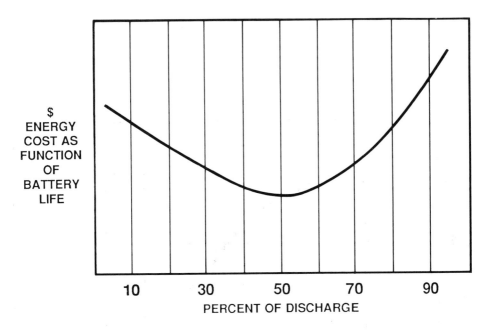

Figure 2.3: Energy Cost versus Depth of Discharge

depth of discharge.

Capacity remaining can be measured by hydrometer tests, or by using a very sensitive voltmeter. In either case, the measurement is not valid if it has been taken on a battery that has been recently charged or discharged.

Figure 2.4 shows how long it takes for acid diffusion to take place. For 10 minutes after a full discharge, the specific gravity remains higher than the actual state of charge. Inside the plate material the acid has been completely depleted. During the next several hours, diffusion starts to lower the acid level outside the plate. About a day later, diffusion starts slowing as SG becomes equalized throughout the battery. Any voltage or SG measurements taken before 24 hours will not be indicative of the actual state of charge.

The hydrometer measures the specific gravity (SG) of the acid. SG is the weight of the acid relative to water. All SG measurements should be temperature compensated ... that is, your hydrome-

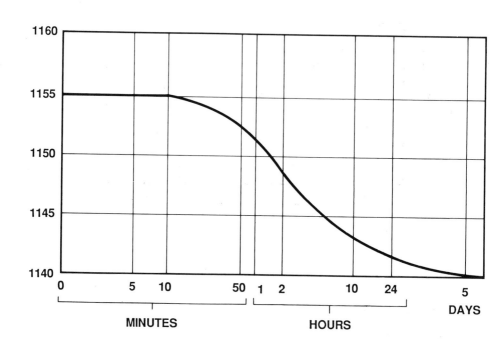

Figure 2.4: Specific Gravity Stabilization Time

ter should have a built-in thermometer so you can add to or subtract from the reading on the float according to the temperature of the acid. Because of the acid diffusion time lag, SG measurements are not valid during charge or discharge. For best results, three identical measurements at half-hour intervals are required during charge.

There is no specific gravity which indicates a full or empty battery. Specific gravity readings are relative, and vary from one battery to the next depending on the initial acidity. You should perform a capacity discharge test and make a record of the results, including SG. Table 2.2, a collection of important values, shows the range of SG typically expected. The higher the initial SG, for instance 1280, the higher will be the SG for a discharged battery. When you buy a new set of batteries, you should find out from the seller what the SG of the electrolyte (sulphuric acid) was prior to putting it into the batteries. What ever it is, (say 1265) then that's what the SG for your new batteries should be when they are fully charged. If the batteries never attain this SG again, suspect that your hydrometer is wrong (possible), or more likely, your charging methods are improper.

The end voltage point of a full discharge at about the 20 hour rate is generally accepted to be 10.5 Volts. You can perform a full discharge by loading your batteries with a known load, and carefully monitor the battery voltage. When it has reached 10.5 Volts, remove the load. Now wait for the SG diffusion to take place and then start measuring SG. After you have attained three identical readings at half hour intervals, record the SG. This is the SG of your batteries at full discharge. By knowing the fully charged SG, and the *empty* SG, simply take the midpoint for the 50% discharged SG. Figure 2.4 shows the stabilization time of SG after a discharge. As shown, after 24 hours (1440 minutes), the SG is approaching its final value, but additional decay takes place for several days.

Battery *cell* voltage is related to SG according to the following rule.

$$Voltage = SG + 0.84$$

Thus, if we determined a midpoint SG of 1.190, then the 50% discharged voltage is shown below.

Item	Value
active materials	lead dioxide, lead, and sulphuric acid
Ah capacity per lb	0.96–1.35
cell voltage	2.12 (12.72); 12–Volt battery
pure DC float voltage	2.17 (13.00)
pulsating float Volts	2.25 (13.5) to 2.27 (13.62)
float current	0.02–0.05% of Ah capacity
gassing voltage	2.40 (14.40)
equalized voltage	2.70 (16.20)
50% discharged	2.03 (12.20); approx.; rested 24 hours
discharged	1.75 (10.50); 5–15% Ah capacity load
charge current	40% of Ah capacity max.; 25% typical
equalization current	7% of Ah capacity max.; 5% typical
Ah efficiency	85–90% (discharge Ah of charge Ah)
storage capacity loss	6–7% per month
temperature/capacity	0.7% loss for each $°F$ below 77$°F$
temperature/SG	0.4 points per $°F$; inverse relationship
temperature/voltage	0.00036 Volts per cell per $°F$; inverse
temperature	120F; maximum
specific gravity	1.250–1.280 full; 1.140–1.170 empty
SG and voltage	voltage per cell $= SG + 0.84$ (rested)
cell to cell SG	30 points needs equalization; 50 replace
height of electrolyte	6mm (1/4 inch) above separators
additives	distilled water only!

Table 2.2: Collection of Important Values

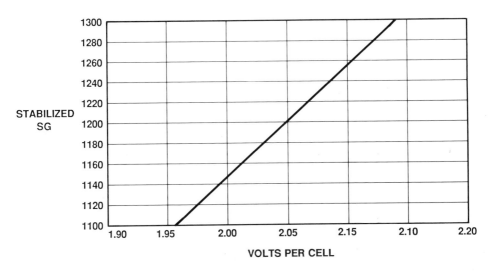

Figure 2.5: Stabilized Open Circuit Cell Voltage

$$(1.190 + .84)6 = 12.18 \text{ Volts}$$

We multiplied by 6 because there are 6 cells in a 12–Volt battery. Note also that the decimal point was re-inserted into the *SG* reading. Figure 2.5 shows the linear relationship between *SG* and open circuit voltage. This figure is generally applicable to deep cycle batteries but will not necessarily be accurate for all batteries.

Voltage measurements, used as an indication of percent capacity remaining, are about as accurate as arthritus and rain prediction ... unless the measurement was made on a battery that has been neither charged nor discharged in the past 24 hours. Even then such voltage measurements may exhibit a 20% error. If you let the battery rest for 5 days before taking the voltage measurement, you may achieve an accuracy of 5%. In short, you can't use a voltmeter in real time as an accurate *fuel* gauge. The red, yellow, and green may look good on the electrical panel, but accurate instrumentation, it isn't. The same goes for those little red lights that *scan* the batteries. Even if you could adjust them to trip at known voltages, they'd be no match for sensitive and accurate digital measurements. Figure 2.6 shows battery voltage versus capacity remaining. Note the broad

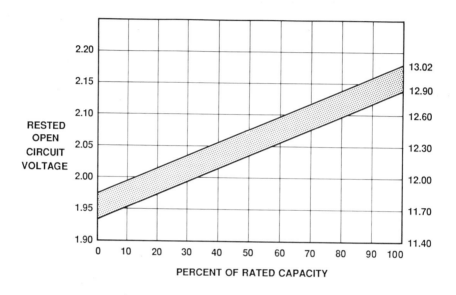

Figure 2.6: Open Circuit Voltage versus Capacity Remaining

range. With the capacity test described in Paragraph 2.5, you can make an accurate chart of your battery.

2.7   Charging Batteries

Since batteries are unique, the best advice you can get concerning the correct way to charge batteries is your battery manufacturer. You must realize though, that the battery manufacturer would rather see a constant 13.8 Volts applied to the battery, than to have the battery charged ignorantly. The common attitude says that all users are idiots who cannot be trusted to manage their equipment. To start a meaningful dialogue with your battery manufacturer, you will need to be armed with the charging information presented here, and demonstrate your ability to follow their recommendations. Remember, the manufacturer is rightly concerned about his product surviving the warranty period. If performance charging techniques are not properly applied,

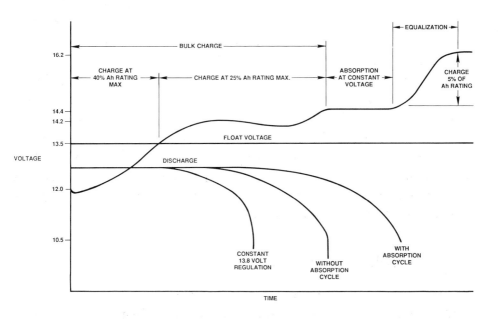

Figure 2.7: The Four Cycles to Charge a Battery

battery life will suffer.

A performance charging system treats the battery to at least three distinct cycles. In technical parlance, the three are known as bulk charge, absorption, and float. For ultimate performance, a fourth, *equalization* cycle should be applied periodically to batteries which have been on a long-term float charge. The four charge cycles are shown in Figure 2.7. Note that current up to 40% of the 20 hour rating can be applied initially. When the voltage reaches about 13.5 Volts, the current should be decreased to 25% or less of the 20 hour *Ah*. Finally, when the voltage reaches 14.2 to 14.4 Volts, it should be held constant, letting the battery absorb current at its own internal diffusion rate. As shown, equalization is the period after absorption. It is not always necessary, and must be performed with a current limited charge source.

Also shown in Figure 2.7 is the relative discharge performances you might expect from batteries charged by different methods. At a constant 13.8 Volts, much of the batteries' capacity is untapped. By charging up to the vigorous gassing voltage, more capacity is attained. For ultimate capacity, an absorption cycle is needed.

It should be noted in the following sections on the distinct charge cycles that the voltages given are for batteries which have undergone enough discharge/charge cycles to reach a stabilized plateau. New batteries will exhibit lower voltages at full charge, and driving them to the full absorption voltage indicated may well harm the battery. New batteries should also be floated at a lesser voltage, and will exhibit less self-discharge current. In general, treat new batteries a little gentle until 25–30 discharge/charge cycles have occurred.

2.7.1   Bulk Charge Cycle

The rate at which a battery absorbs charge cannot be easily calculated. The rate is primarily dependent on the diffusion of electrolyte throughout the active material in the plates. The thicker the plates, the slower the diffusion, and hence the less the charge acceptance. Thick plates are required for sustained discharges and long battery life, so it isn't possible to design a (conventional) deep cycle lead acid battery that can be *fast* charged on a regular basis.

A discharged battery can absorb a heavy *bulk* charge. This heavy charge can be applied to the battery until such time as the acid begins to *gas* (bubble). At about 14.4 Volts (at $77°F (25°C)$), a portion of the electrolyte will separate into hydrogen and oxygen. At this point, current through the battery must be reduced to prevent excessive loss of acid, and also to prevent a rapid build up of heat. Gassing can be observed in a battery: wear eye protection to avoid acid splash.

The amount of electrical capacity that can be delivered to the battery prior to when gassing commences depends on the charge current, and the plate thickness. For relatively thin plates, a current of 40–50% of capacity, will deliver about 50% of full charge by the time the gas point voltage is reached. A 75% charge can be delivered at a charging current of about 20–25% of the battery capacity ... 20–25 Amperes for a 100 Ah battery. If you want to deliver 85% capacity before gassing, you must limit current to about 10% of the Ah rating of the battery. These numbers are all lower for the older design, thicker plate deep cycle batteries. Such batteries should never be subjected to more than a 25% charge current, and should not be charged to 14.4 Volts.

(An important detail should be noted here. Alternator regulators for automotive application are set so that they drive the battery up to about 13.8 Volts. At this voltage, some gassing occurs, but not the vigorous kind that is indicative of performance charging. Since a battery cannot be maintained in a state of vigorous gassing too long, 13.8 Volts is a safe compromise where performance is not an issue. In any system greater than *spartan conditions*, performance is very much an issue. The standard regulator, under the banner of *cheap to make* falls short of providing the battery with a full charge. Special purpose regulators are needed for proper deep cycle battery care, and because they are not mass produced they naturally cost more.)

The objective of the bulk charge cycle is to deliver as much charge as possible in the shortest time. Temperature limitations for both alternator and battery must be observed, and as noted earlier, $120°F$ $(50°C)$ is the maximum for the battery. Now, 400–500 Ah of battery capacity is not unusual for a cruising system. To charge at 20–25% of this will require an alternator capable of sustained output of 80–125 amps. To deliver a 40% bulk charge the alternator should be able to supply an initial current of 160–200 Amps.

Charging at a slower rate is not without its merits. Diffusion of acid can more readily keep up with the current, so that when the gassing point is reached, more total charge will be delivered. Temperature buildup isn't as rapid, and the detrimental effect of heat on the plates and separators is minimized.

A charging rate of 10–15% of capacity up to 14.4 Volts is a safe but not a high performance regimen. If you choose to charge at a greater rate, beware of high temperature and apply temperature correction to the gas point voltage of 14.4 Volts at $77°F(25°C)$. Refer to Table 2.3 for temperature/voltage compensation data. Electrolyte Temperature reference for values in text is $77°F$.

Charging a battery requires that you put in about 15% more Ah than you took out. That means a charge rate of 10% will take 11.5

Temperature $°F/°C$	50% Discharge (Volts)	Float (Volts)	Gas Point (Volts)	Equalized (Volts)
37/3	12.29	13.59	14.49	16.29
47/8	12.26	13.57	14.47	16.27
57/14	12.24	13.54	14.44	16.24
67/19	12.22	13.52	14.42	16.22
77*/25	12.20	13.50	14.40	16.20
87/30	12.18	13.48	14.38	16.18
97/36	12.16	13.46	14.36	16.16
107/42	12.14	13.44	14.33	16.14
117/48	12.11	13.41	14.31	16.11

Table 2.3: Temperature/Voltage Compensation Table

hours if you started out with a flat battery. It only makes sense to charge at a faster rate until the gas point is reached, but charge acceptance will be less so you can't halve charge time by doubling charge rate. For the longest life, keep it cool.

During charge, there is always some parasitic oxygen generated at the positive plate. This requires some amount of energy, which is the 15% which must be put back into the battery above and beyond that which was discharged. As the battery approaches a full charge, more and more of the charge current goes into oxygen generation. Charging is basically an oxidation process. Once most of the active material in the positive plate has been converted, the current carrying grid tends to oxidize. With sufficient overcharge, the positive plate will *corrode* until the battery completely fails. Because a full charge is not achieved abruptly, knowing when to stop charging is difficult. Some amount of overcharge is required to assure a full charge, but it must be limited to prevent damage to the positive plate. You should note that corrosion of the positive plate is worse for deep cycle batteries which use antimony in the positive plate.

2.7.2 Absorption Cycle

As we explained above, the battery is a long way from fully charged once the gas point voltage is reached. At a high rate of charge, only 50–60% capacity is restored. Thus, charging should continue, but the rate of charge should be lessened. By maintaining a constant voltage, the battery will absorb charge at its own natural absorption rate . Any charge current that is greater than the natural absorption rate only serves to heat the battery and reduce its life. Two battery parameters are of concern here ... the constant voltage to apply, and the natural absorption rate. The natural absorption rate of the conventional lead acid battery is very much a function of battery design, and the state of charge. After delivery of a bulk charge, several more hours of absorption charging may be required to attain full charge, due to the low natural absorption rate at near full charge.

The constant absorption voltage that many battery manufacturers recommend is 13.8 Volts, the same voltage they recommend for charging. This explains the prevalence of constant voltage chargers set at a nominal 13.8 Volts. This value is actually an all around compromise. If you have the option of adjusting your charger, then you may elect to charge at 20% of *Ah* current until the battery reaches about 14.4 Volts. Now switch to voltage regulation at this voltage value, and let the battery absorb at its natural rate until the charge current has fallen to 1 or 2% of its *Ah* rating. The charger should now be switched to float regulation. Because it takes a long time for the battery current to fall to 1%, you may elect to terminate the charge at 5% of *Ah* rating. An old battery may never get to a current as low as 1% of its *Ah* rating.

You should note that chargers which automatically switch from charge to float when the gas point voltage is reached do not properly apply this absorption cycle, and thus fall short of fully charging the battery.

There are some precautions about absorption at a gas point voltage. Both hydrogen and oxygen are being produced. If enough accumulates, an explosion can result. The oxygen rich environment can hasten corrosion in nearby electrical equipment. Also, see our advice below regarding gas recombinant caps, which can overheat and fail.

2.7.3   Float Cycle

A fully charged battery can be *floated* for indefinite, but extended periods of time (years) with no loss in capacity or life expectancy. Floating is the term given to the following circumstances ... a constant and closely regulated voltage source is applied to the battery, with the voltage setpoint being very close to the open circuit potential of the battery. A proper float voltage under these conditions is about 15–20 millivolts per cell above the open circuit voltage. For a 12–Volt battery, this results in a float voltage of 13.02.

These are all nice facts to know, but the facts must be used within the context to which they apply. There are manufacturers who sell equipment to float batteries, even quoting the correct number of 13.02 Volts. So what is wrong? First, and foremost is the circuit topology generally used for battery chargers. At very best, they output *DC* current pulses, requiring the battery itself to average the pulses. The chargers generate pulses because no capacitor filters are used after the rectifier diodes, as is the case for actual *DC* voltage supplies. An *AC* line voltage charger delivers current pulses at twice the line frequency (120 times a second in the U.S.). An alternator delivers pulses at the rotational *RPM*, times the number of armature poles. The net result is high current charge pulses followed by blank periods when no current can be delivered.

If you aren't using the batteries during the time they are maintained on float cycle, it isn't as detrimental to use 13.02 Volts as the case when the batteries are in use. When the battery is being charged under a load, then during the blank periods when the charger cannot deliver current, the battery is being discharged! It doesn't take much of an average current to keep the average voltage at 13 Volts, but the net result is a slowly discharging battery. If left in such a float cycle long enough, the battery can become discharged to the point where sulfation occurs, the natural result of leaving a discharged battery standing. Under the circumstances of a low float voltage and pulsed current, the life of the battery cannot be accurately estimated due to the irregular discharges and complicated *quasi float* current pulses. In general, you can expect less service life from such a low float voltage than one which is higher, and thus more capable of restoring the lost capacity which occurs during blank charge periods.

For 13.02 Volts to be a correct float voltage a pure DC charger must be used, with the maximum pulsations in output to cause no more than 180 millivolts of ripple for a 12–Volt battery. Furthermore, the pure DC supply must be capable of providing 100% of the load demands, thus preventing even small discharges from occurring. Only under these circumstances can *floating be free*. As we noted above, floating at 13.02 Volts with a pulsating type charger will be detrimental to the life of the battery, allowing some of the active material to sulfate.

The objective in floating is to maintain a full charge, applying no more current than necessary to do so. In this regard, when we take into consideration the circuit topology of chargers, and the fact that the batteries are being used during float, and sometimes at a greater rate than the charger can supply, what is the proper float voltage? It turns out that batteries are best floated with a small trickle current, rather than a constant voltage. If you have a regulator that can regulate to a sensed current, then sense battery current and adjust the charger to deliver about 0.05% of its capacity, (50 milli-amps for a 100 Ah battery). New batteries should be adjusted for about half this value, and old batteries may require twice as much to maintain full charge. Temperature must be considered as usual. Above $77°F(25°C)$, self-discharge current doubles for every $18°F(10°C)$ rise in temperature.

To determine the correct float current for your batteries, apply enough current to maintain the battery voltage between 13.2 and 13.5 Volts (at $77°F(25°C)$). Remember, a properly adjusted trickle charge is better than a constant float voltage.

If you must make do with only voltage regulation of a normal pulsating charger, then under most circumstances, an average voltage of 13.65 will suffice. This is just high enough to restore the discharges made during the blank periods when the charger is unable to output, yet it is low enough so that excessive current is not passed through the batteries. For warmer weather, reduce the float voltage to 13.5 Volts.

2.7.4   Equalization Cycle

Most of the charge is delivered during the bulk charge cycle and if followed by an absorption cycle, the battery will be 90–95% charged. Getting the extra few percent available to you is just not worth the effort of equalization on a regular basis.

Occasionally, however, the battery should be driven beyond the gas point with a small constant current, until the battery reaches its highest natural voltage. Such an operation is called equalization, because it *equalizes the specific gravity* within each cell, and from cell to cell. Be advised that at least one battery charger manufacturer uses the term *equalize* to mean the bulk charge cycle. Such mis-use of terms runs counter to established practice and adds to the confusion that exists about charging. Another manufacturer has adopted the terms *condition* and *conserve* to mean equalize and float.

Equalization is periodically required because of the small differences between cell construction and specific gravity. Any differences in the charge or discharge currents which may occur from cell to cell aggravate the cell differences until equalization is required. How, you ask can charge and discharge currents for the cells be different? Aren't the cells connected in series? Of course, but dirty battery tops take their toll, slowly, but continually. Furthermore, it isn't uncommon to find 12–Volt batteries made up of two 6 volt units, physically isolated, one running cool and clean, the other running hot and dirty. While this may not be avoidable, you can expect the hot and dirty battery to require more care. Most certainly, it will age faster, aggravating any differences between it and the other which existed to begin with. Remember too that each cell is in a separate bath of acid with slightly different active material. The cells cannot be expected to discharge and charge exactly the same.

To properly equalize a battery requires that each cell be equalized independently! That means you are dealing with six, 2–Volt cells, not the one 12–Volt battery. Independent equalization is difficult to do, and if you are lucky enough to have 12–Volt batteries with internal connections between cells, you can't perform independent equalization. (The need to treat each cell independently is much less with internally connected cells because all cells are operated in the same clean local environment.)

If you are still determined to properly equalize, Figure 2.8 shows how it is done. Equal resistors with 1% or better accuracy are placed between each 2–Volt cell, providing a parallel current path, and allowing each cell to follow its own natural rate. The resistors should be such that they conduct 1% per *Ah* of the battery rating ... i.e. for a 100 *Ah* battery of six cells, you need 6 resistors of 2 ohms, 3 watt rating, 1% tolerance. Don't forget to disconnect this equalization network after use, otherwise it will discharge the battery. Note that series connected 6–Volt batteries should be equalized with a resistive divider. If you can't make the 2–Volt connections, use two resistors instead of 6, and tie the center of the two resistors to the center tap of the two 6–Volt batteries. The two resistors should still conduct current equal to 1% per Ah as above, and should have a 1% accuracy tolerance as well.

There are other problems with equalization which you should be aware of.

First, the current to be applied should not be over 7% of the *Ah* rating of the battery. A lesser number of 3 to 5% is safer. The equalization cycle should never last beyond 4 hours, even if the voltage doesn't reach 16.2 Volts.

Secondly, special warning is given to those who are using the platinum catalyst caps which recombine hydrogen and oxygen. These caps are not designed to recombine large amounts of gas. A 5 Ampere current will generally push them to their limits ... producing enough heat to melt the caps. After that point, hydrogen and oxygen must be released into the atmosphere, through a small vent in the now very hot cap. Explosion may result. Unfortunately, when you need the caps the most, we recommend that you remove them for safety reasons.

(We question the wisdom of using gas recombinant caps. First, they don't work when you need them, which is any time that you are charging above the gas point with 5+ Amps. Even the caps which appear to work are not 100% efficient, so under a performance charging regimen, there will still be hydrogen and oxygen escaping. In the meantime, you are lulled into a false sense of security. If a cap should fail, and you have not been checking the water loss, battery damage

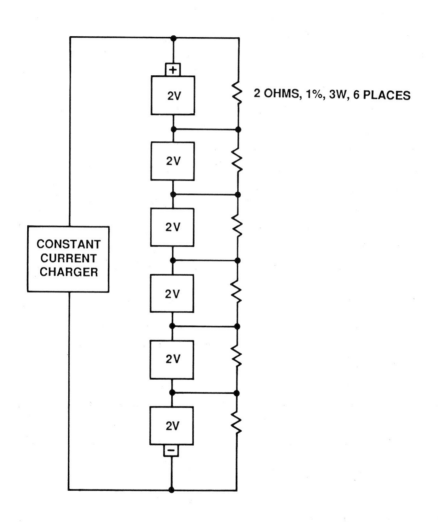

Figure 2.8: Equalization Network for 12–Volt 100Ah Battery

will result. The people that sell gas recombinant caps *should* be aware of these problems with the product and *should* be sure the customer is aware of the caps' shortcomings. The bottom line is, be a wise consumer, ask questions. If you should have an explosion at sea, then it's too late. If in fact the caps performed as we are led to believe, then there would be extensive specifications on their performance. Without guaranteed limits of performance which apply to the caps, we recommend that you *do not* use them. What you need instead, is ventilation, and a regular program of battery maintenance.)

Given all the hassles of equalization, is it really worth it? If you are using cheap batteries, i. e. in the $150 range for an 8D, the answer is a definite NO. If you have purchased an expensive set of deep cycle batteries you will want to do whatever you can to extend their life.

Remember, a battery is never 100% charged unless it is equalized by a low constant current and its terminal voltage is allowed to rise to the battery's highest natural voltage of about 16.2 Volts. After conditioning your batteries, let the batteries rest 24 hours or longer, and before using them, measure the *SG* and the voltage. These measurement readings are the *full* mark for your batteries.

2.8 Series and Parallel Combinations

System capacity can be increased by using series and parallel combinations of batteries. Two 6–Volt batteries connected in series make a 12–Volt battery, with an *Ah* capacity equal to the capacity of the smallest 6–Volt unit. Connecting two 12–Volt batteries in parallel still results in a 12–Volt system, but now the capacity is equal to the sum of the individual capacities.

Either method is appropriate under given conditions. Foremost of these conditions is that the batteries be new when first connected, and they always be operated as first connected. Putting an old battery with a new one is a poor practice, whether in series or parallel. Series connected batteries should be as close together as physically possible, since the weaker of the two sets the overall capacity, and any differences in operating environment can adversely affect the other. Series connected batteries should be equalized with the resistive network that was described above. Failure to do so will result in the stronger battery being overcharged, and the weaker of the two being undercharged.

Batteries of the same capacity from the same manufacturer are to be preferred over unequal capacities and different manufacturers.

Another condition of series connected batteries is that deep discharges be avoided ... the discharge current of the highest capacity cell has the effect of a charge current of reversed polarity on the weaker cell, further weakening it. If you are using series connected batteries, limit discharges to about 60% of capacity. That means normal service will be in the range of 85% down to 40% at the most ... you can only use about 45% of the rated capacity.

Before connecting 6–Volt units in series to make a 12–Volt battery, the 6–Volt batteries should receive a full charge in parallel as 6–Volt batteries. The 6–Volt batteries should also get a full equalization charge before going into service as a 12–Volt bank.

In general, best performance will be achieved by parallel connections, thus series connections should be avoided unless space availability and battery availability force you into building extra capacity from 6–Volt units.

Parallel batteries supply the sum of the individual capacities. As long as batteries are new to start with, unequal sizes can be connected in parallel. Where you might not have room for two 200 *Ah* batteries, one 200 *Ah* unit and one 100 *Ah* unit could be paralleled to yield 300 *Ah*. The batteries will share discharge and charge currents according to their individual capacities. While deep discharges are to be avoided, parallel units do not have the reverse polarity charge effect mentioned above for series connected batteries.

Whether you have one battery or a hundred, and whether they are made up of series, parallel or series-parallel combinations, you should have accurate instrumentation and you should keep a log of values. Any sudden or significant change in voltage or charge current is a symptom of impending failure. When a cell fails in a series connected system, the cell acts as a high internal resistance. The batteries will neither discharge or charge with typical high currents. When a cell fails in a parallel connected bank, charge currents may be high, but the voltage won't rise to its customary value. By accurately recording typical discharge and charge conditions, spotting an impending failure can be done before disaster strikes.

Cell failures are extremely rare in batteries which have been maintained properly. As noted above, equalization can restore cell to cell balance when *SG* indicates the need. As long as you don't kill your batteries by excessive discharges and overcharges, expect your batteries in series or parallel to survive until normal wear reduces capacity to 80% of its new rating.

2.9 Safety Precautions

The electrolyte of the battery, since it contains sulfuric acid, can cause blindness if splashed into the eyes. Ingestion of electrolyte can cause death. Simple exposure to skin can cause painful burns. Protect yourself when measuring battery *SG*, or otherwise handling the acid. Don't use containers for the acid which are typically used for other uses. We know one person who nearly died from swallowing acid left in a cup normally used for drinking water. By all means, mount your batteries in such a way that inquisitive children cannot perform experiments.

By now, you know that batteries conduct electricity. Such things as rings, watches and wrenches make excellent conductors, and the average deep cycle battery has enough energy to melt them into a bubbling puddle of scrap. Take off your jewelery before working around any electricity ... it can be replaced if lost. A finger, amputated by the burn method is not so easily replaced. Use extreme caution when working around batteries with any tools. Be positive that batteries are protected from falling objects like screws, nails, wrenches, etc. Molten sparks from such conductors can fly a long ways, and eyes seem to be magnets for such dangers.

A battery charged to the gassing point produces both hydrogen and oxygen. Hydrogen is highly explosive, and oxygen promotes combustion. Make sure that you have ventilated battery compartments which are clear of any open flames or sources of sparks. Batteries can explode when charging if any form of visible heat is brought within proximity to the hydrogen. Don't rely on gas recombinant caps to work. As noted above, they are not rated for much gassing in any case. You need ventilation.

Batteries which have antimony in their plates, (most, if not all deep cycle batteries) release two gasses, arsine and stabine, which are health hazards when inhaled. Again, ventilation is needed.

We noted above that an oxygen rich environment can speed up the normal corrosion that occurs in electrical equipment. We examined a relay once which was used for an autopilot. The relay had been mounted a few inches above one of the large batteries. Inside the relay case was a smokey haze, part corrosion, but also part residue from burning oxygen. The alternator which charged the batteries had an unusually high setpoint about 14.4 Volts, so gassing was continuous whenever the engine was running. We can only wonder how much flame jumped the gap between relay contacts each time the autopilot reversed its drive motor.

Incidentally, the boat had recently been surveyed by a marine surveyor and inspected by the Coast Guard. Neither had questioned the safety of the battery and auto pilot position.

2.10   Maintenance

A very important aspect of maintenance is record keeping. As we indicated earlier, you should know full charge SG, and end-point SG. From these two, you can calculate the midpoint or 50% discharge SG. Keep a record for each cell in your batteries.

By all means, keep the electrolyte above the separators. (The separators protrude slightly above the active plate material, both being visible by removing the cell cap.) Electrolyte should cover the separators by about 1/4 inch, (6mm). Exposing the plates to air is a fatal blow to the batteries. That portion of the plate that drys out in the air is permanently out of action, even if covered with electrolyte again. On the other hand, if you overfill the cells with water, you dilute the acid, and will not get the full capacity from the batteries. Always use distilled water to replace that which is lost due to gassing. Never add acid ... that will worsen self-discharge and increase corrosion of the plates. Don't even *think* about curing battery ills with patent additives. (Most batteries will tolerate low mineral water, but why risk damage when distilled water is available.)

Keep the tops of your batteries clean! Dirty batteries self-discharge much faster than do clean ones. The first step in preventing dirty batteries is to keep the acid in the battery. Don't charge them with the caps removed such that an acid mist is sprayed over the top. When you remove the gas recombinant caps for absorption or equalization, place a loose fitting bundle of paper towels over the hole to absorb acid mist. Wash the battery tops with clean water afterwards.

Salt water is an electrolyte, and in a damp bilge, even batteries that have not been treated to an acid bath, can start to corrode around the terminals. The solution for this problem is to apply petroleum jelly to the terminals (with the battery cables connected) before corrosion starts. If you already have corrosion, clean the terminals with baking soda and water. After drying thoroughly, apply the petroleum jelly.

There are special chemicals sold in automotive stores for corrosion problems. Save your money ... those products are sold to people who don't know that inexpensive household products do a better job.

Keep your batteries cool and dry.

If your batteries are no longer needed on a regular basis, charge them with a full equalization charge before storing them. Wipe them clean, and apply petroleum jelly to the terminals. You should give the batteries an equalization charge on a monthly basis while in storage. If they go three months without an equalization charge, expect a permanent loss of capacity.

While lead sulfate is a normal by-product of discharge, the sulfate crystals that form during self-discharge are larger. These crystals completely surround clumps of active material and insulate the clumps from the plate grid, removing otherwise good material from service. If a battery self-discharges below 10.5 Volts, it will never fully recover.

Whenever a battery is too deeply discharged, the sulfate crystals become slightly soluble in the electrolyte. This problem is exaggerated for sulfate which is formed due to self-discharge. The dissolved sulfate diffuses through the plate separators forming dendrites which may well short the plate. Dissolved sulfate also tends to settle into the bottom of the cell, and will short the plates when enough accumulates.

Deep discharges are never advised ... deep self-discharges are extemely destructive.

2.11   How *not* to care for Batteries.

High quality batteries are amazingly resilient, but a determined killer can always murder one. Here are a few ways to do so.

- Discharge the battery flat dead and then leave it that way for awhile. The sulfated plates will crystalize so that no charger in the world can recover the capacity.

- Freeze the battery now that it is flat discharged. This leads to speedy death since the low SG of the electrolyte will allow it to be frozen like water.

- Leave the battery connected to a constant voltage charger that has an output voltage greater than 13.8 Volts. It will keep pumping current through the battery until it finally corrodes away the positive plate.

- Partially charge the battery using a constant voltage charger with an output less than 14.4 Volts, or one which automatically switches to a low float voltage as soon as it reaches the gas point voltage. By never allowing the battery to reach the gassing voltage, or automatically tripping to a low voltage when it does, the tops of the plates will become permanently sulfated ... out of service.

- On a regular basis, discharge the battery well below the optimum 50% capacity point.

- Operate the battery above $120°F(50°C)$. If you really have a mean streak, add a little acid to raise the SG before overheating the battery.

- Let the electrolyte get below the tops of the plates so they dry out. The lower you let the electrolyte get, the more effective is your torture.

- When the battery is not in active use, let it self-discharge instead of floating it at a pure DC voltage of 13.02 Volts.

- Above the gassing point of 14.4 Volts, continue to charge it at more than 7% of it *Ah* capacity. This will hasten the corrosion on the positive plate, and with any luck you can drop the whole plate into the bottom of the battery.

- Periodically, add acid or dirty water.

- Fry your alternator until a diode shorts and then wonder why your alternator ammeter seems to indicate *AC* going to the battery (it is).

- While in normal *liveaboard* circumstances, set your pulsating type charger to a low voltage such as the recommended pure *DC* float voltage of 13.02. Now continue to use your normal appliances until the *SG* of the battery is hovering around the 20% capacity range. Now take your boat out for a slosh so the dissolved sulfate falls to the bottom of the battery.

2.12 How to use Batteries

We noted above that batteries provide more capacity when discharged intermittently, rather than continuously. Also noted was the fact that voltage measurements are not a valid indication of capacity remaining unless the battery has been rested for at least 24 hours. Even after 24 hours, an inaccuracy of 20% exists. A longer rest time is required on thick plate batteries. Finally we noted that a 50% discharge yields the lowest energy cost, thus it is to your advantage to limit discharges to the 50–60% range.

Putting all these facts into one *scenario*, leads to this recommendation. Use two equal sized battery banks. Alternate between them on a daily basis. Just before switching banks, measure the voltage of the rested battery and use the reading to determine capacity remaining. When both batteries reach 50% discharged, it is time to recharge. As you can see in Figure 2.3, 50% is not a bewitching number. Greater discharges, however will reduce battery life. The key is to discharge about 50–60%.

To manage intermittent use as just described, enough *Ah* capacity for several days is required. The issue of determining how much capacity you need is deferred until later in the book.

Long slow charges are better for long life than fast charges, but even fast charges can be done as long as battery temperature is not allowed to rise too high. As the voltage on the battery approaches 14 Volts, be sure that current has fallen to the 10–15% range. Avoid large constant current charges that push all the way to 14.4 Volts, even if the current is shut down at that point. It is easier on the batteries and the battery can be charged fullest by applying an absorption charge at a constant voltage. The battery accepts current at its own natural rate in the absorption cycle, avoiding parasitic oxygen generation and the resulting oxidation of the positive plate.

Putting a full charge on the batteries is not necessarily required each time you charge them, although you should always apply some of the absorption cycle. During absorption, rather than wait for current to decrease all the way to 1% of the Ah rating, settle for 5–10%. You will be giving up some capacity, but won't be wasting alot of time for a few percentage points. Periodically, however, perform at least a full absorption cycle. If you are a purist, you will follow with an equalization cycle.

Properly cared for, batteries will last for many years. Figure 2.9 shows the number of discharge cycles versus depth of discharge that you might expect of a high quality deep cycle battery. These curves reflect optimal discharge and charge conditions. Without suitable instrumentation and control, you most certainly will be in the low range.

2.13 Summary

The key to performance is always knowledge about operating characteristics. In this chapter, we have presented much useful information about batteries in general. As noted, batteries are unique, and the specific recommendations of your battery manufacturer should be adhered to. Without accurate instrumentation and a disciplined operation and maintenance program, achieving performance is unlikely.

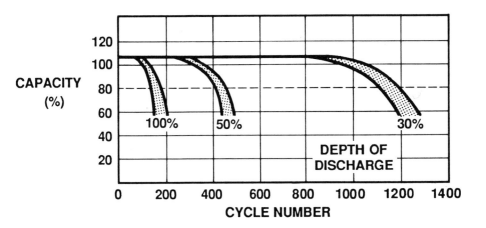

Figure 2.9: Cyclic Life versus Depth of Discharge

Chapter 3

Immobilized Electrolyte Batteries

3.1   General Information

Patents covering the first sealed lead acid battery were issued to Adolf Dassler, of Germany, in 1933. In 1951, U.S. patents were issued. Continual improvements have been made since then, and use of sealed batteries has risen accordingly. Early users of sealed lead acid batteries took advantage of the position independence characteristics of the batteries, an advantage previously held by Ni-Cad[1]. Sealed lead acid batteries offer more capacity less expensively than Ni-Cad. Though still expensive, the sealed units find major applications in Uninterruptible Power Supplies (*UPS*) that protect sensitive and critical electronic systems from catastrophic power outages. Emergency lighting systems now use sealed batteries almost exclusively. Sealed batteries can be used in the home and office environment where gasses and acid fumes from a conventional battery would be objectionable and unsafe.

In recent years as many as 60 manufacturers have joined the battle for market share, and consequently, performance has grown while prices have actually decreased. Sealed technology may not be so economical in the near future that it finds widespread use under the hood

---

[1]Nickel cadmium batteries have been made commercially in the United States since 1909.

of automobiles, but anybody seriously thinking about a performance electrical system must strongly consider the many advantages offered by today's sealed lead acid batteries. Be advised, however, that not all of today's sealed batteries are appropriate for a cycling application. Many units are designed for standby float service, and will not perform well if cycled through deep discharges.

3.2   Nostalgia

We have a friend who owns a 1941 Chevrolet pickup. He believes that truck quality has deteriorated ever since. We understand nostalgia, and won't argue that his truck was built well, but he has no real interest in performance. He's learned how to shift the square cut gears in the transmission, and 55 *MPH* is now a goal! If you find nostalgia in batteries with hard brittle cases made today with the same tugboat technology used forty years ago, and are willing to settle for performance levels accordingly, then don't read any further.

It used to be that regular maintenance was required on lead acid batteries. Acid level had to be checked frequently, and topped off with distilled water. Self-discharge rates were high enough so that even batteries not in service had to be frequently charged or suffer permanent sulfation. Batteries had to be operated in an upright position[2]. Extremely deep discharges had to be avoided and recharge was required immediately to prevent sulfation. Overcharging was such a problem that users put up with time consuming low voltage chargers. Even the technically mindful, using *fast* charging methods had 55 *MPH* as a goal.

3.3   Sealed Technology

The original sealed lead acid battery captivated the electrolyte in a gel. Recent methods coat thin, closely separated plates and the fiberglass and felt mat separators with just enough electrolyte to *dampen*

---

[2]Long upwind passages on the same tack actually expose part of the battery plates to air with permanent loss in capacity the consequence.

them. This variety of sealed batteries is known as *absorbed electrolyte* batteries.

Special micro-porous fiberglass separators are used. The plates and separators are compressed when assembled, giving the sealed battery great resistance to vibration. Compression also reduces the electrical resistance of the battery, allowing it to generate large amounts of current, such as required when starting big diesels. Of the competing technologies, the gelled unit appears to support faster charges than the absorbed electrolyte version. For absolute performance, the gelled unit should be chosen since it stands up to faster discharges and may be left for short periods in a discharged state without permanent damage. The absorbed electrolyte battery needs to be treated with about the same care as a conventional wet battery.

An old fashioned battery requires thick plates to provide deep cycle performance. The principal reason for this is the fact that the supporting plate grids must be made thick to provide mechanical strength, and to tolerate the corrosion which occurs on overcharges. The plates in a conventional battery are suspended in fluid, not compressed with the separators. Conventional grids even use a lead/antimony alloy to gain strength even though the antimony contributes to self-discharge, hydrogen generation before full charge can occur, and internal corrosion.

The thicker plates of conventional batteries cause problems with acid diffusion ... fast discharges and charges are ruled out by design. Using a thick plate conventional deep cycle battery for large starter currents can result in poor engine cranking. Larger boats that use electric windlasses or electric winches are forced to buy extra battery capacity to get enough high current capability to operate properly.

Sealed units, on the other hand, are constructed with thin, pure lead or lead-calcium grids which are supported by fiberglass separators and tough, strong (usually polypropolene) cases. Some manufacturers use a lead-calcium-tin alloy for the plate grids.

Although the chemical reactions inside the sealed battery are the same as that of the conventional unit, the high purity of the material minimizes hydrogen generation during overcharge conditions. The active materials in the battery are balanced so that the positive plate will become fully charged and begin evolving oxygen before the negative plate is fully charged. To achieve this, more active material than strictly necessary is put into the negative plate. Under normal operation the negative plate is never fully charged and consequently no hydrogen gas is generated.

Because the plates are closely spaced and not separated by liquid electrolyte, homogeneous gas transfer occurs, resulting in recombination of oxygen inside the cell. Internal pressures of $1 - 4\Psi^3$ contribute to oxygen recombination. Efficient recombination occurs at $1 - 2\Psi$. As a safety measure, pressure release valves (Bunsen type) are fitted so that continuous high overcharges do not cause the case to rupture. These valves are set to release gas in the $5 - 10\Psi$ range.

While the principles of sealed batteries are easy enough to understand, development of ways to immobilize the electrolyte and allow gas transfer to occur required much effort. As noted above, there are two types of sealed batteries, gelled electrolyte and absorbed electrolyte.

3.4 Sealed Advantages

Besides being sealed, and never needing maintenance, the sealed battery can be operated in any position. If operated upside down, precautions should be taken to leave clearance around the safety release valves.

Overcharges are *not* required to fully charge, and the natural absorption rate under a combination charge, absorb, and float voltage (13.8 Volts) is about twice that of a wet battery.

Many thin plates with a correspondingly large surface area support rapid discharges and charges, since acid diffusion through the thin plates is more rapid. Because the batteries are designed for high current, no current limiting is required for charging. Under properly controlled conditions some sealed batteries can be fully charged in about half an hour. To accomplish this on a 100 Ah unit, you will need a source capable of delivering 600 amps while maintaining an output voltage of 14.1 volts (plus or minus 1%). This puts those 200 Amp alternators in the wimpy class!

---

[3] The Greek letter Ψ is used to represent *pounds per square inch*.

With high purity lead grids, corrosion at the positive plate, the usual battery killer, is practically eliminated, assuring a long life. At 80% depth of discharge on a daily basis, with proper charge control, you might expect 200-600 cycles. At the optimum 50% discharge level, 400-1000 cycles are possible. With 25% depth of discharge, life may exceed 2000 cycles. At a 10% depth of discharge, expect 20,000 cycles. These numbers compare favorably with even the best of conventional deep cycle batteries. In practice, while the conventional battery may offer more deep discharge cycles if treated carefully, the sealed battery stands up better under tough conditons such as fast charges and very deep discharges. Because of this fact, you may achieve both higher performance and more cycles than a conventional battery.

The sealed battery is not as sensitive to depth of discharge as its predecessor. While 50% discharge will still result in the most power for your money, it isn't always convenient to recharge at the 50% point. A high quality sealed unit meeting European standard DIN 43539, must withstand 100% discharge followed by 30 days connection to the load, and suffer no permanent damage! After such abuse, it must reach full charge at 13.8 volts within 48 hours. While it may loose 25% capacity on the first charge, it is expected to regain full capacity on subsequent cycles, as well as its naturally high absorption rate. No wet battery made can stand up to this kind of torture.

It should be pointed out that not many sealed batteries will withstand the torture test of DIN 43539. Even those that do may require some special treatment to bring them back. Toward the end of this section is a synopsis of our own experience with this subject.

Some sealed units have very low self-discharge rates. Units not in service should be charged every 16 months. In colder climates the battery only needs attention every couple of years. Here's a battery that can be left uncared for over the winter as long as it is fully charged prior to storage!

Old fashioned thick plate batteries are limited in current capacity by acid diffusion. Because sealed batteries do not suffer this problem, more capacity is available under heavy loads such as those encountered when operating inverters to power microwave ovens or electric winches. The output voltage under such heavy current stays higher, keeping the efficiency of the inverter higher, resulting in more usable

State of Charge (%)	Rested Voltage (48 hours)
100	12.80
75	12.60
50	12.40
25	12.20
0	12.00

Table 3.1: Capacity Remaining vs. Voltage

power for your money.

3.5   Determining Capacity Remaining

Because you can't measure the specific gravity, (SG) of a sealed battery you must rely on a voltage measurement to determine the capacity remaining. The electrolyte in the sealed battery takes longer to stabilize after a charge or discharge ... in some cases twice as long. Whereas we recommended alternating between conventional batteries on a daily basis, a user of sealed batteries should alternate every other day for best accuracy. Before switching from one bank to the other, measure the rested battery voltage. Given in Table 3.1 is the state of charge versus depth of discharge for a sealed battery from Ample Power Company[4]. While all batteries are not the same, Table 3.1 can be used as a general guideline.

3.6   Charging

As we noted earlier, sealed batteries can be rapidly charged. At a constant voltage of 13.8 volts, natural absorption rate is about 50% of capacity. This rate decreases as the battery approaches the full charge condition, but full charge on a flat dead unit can be accomplished in about 3–4 hours, by applying a constant 13.8 volts. Figure 3.1

---

[4]Catalog #1017.

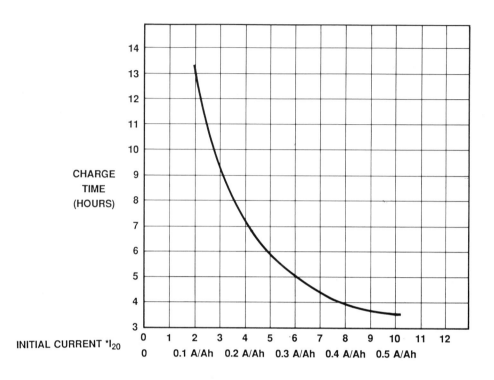

Figure 3.1: Time to Charge versus Initial Charge Rate

demonstrates the time required to reach a 90% charge with different initial charge currents. (Catalog #1017 from Ample Power Company.) At the start of charging, the battery is 100% discharged. As shown, an initial charge current of 50% capacity will charge the battery in about 3.5 hours.

To develop the chart, a fixed voltage of 13.8 Volts was applied to the battery, but limited in current delivery. The numbers of 0–12 along the bottom are multiples of the 20 hour capacity. A unit rated at 100 Ah for 20 hours will have a rate of 5A. At 2 times that rate, about 13 hours will be required to recharge. At 10 times the 20 hour rate, a full charge can be attained in about 3.5 hours.

Perhaps 3.5 hours is acceptable, but faster charges can be obtained by a higher charge voltage. Raising the voltage to 14.1 volts permits initial charging as fast as current can be delivered. Because of the highly pure materials, leaving the battery for as long as a week at 14.1 volts will not substantially damage it, though it is recommended that fully charged batteries be floated at 13.2-13.80 volts.

Figure 3.2 shows the proper range of float voltage versus temperature for sealed batteries. For the longest life, always use the lower of the values. Figure 3.3 shows the float current of a 100 Ah battery at three different float voltages. The effect of high voltage and high temperature is dramatic, indicating the necessity of a properly adjusted charger. Note that between 13.2 and 13.5 Volts, there is little difference in current as temperature rises to the upper limit of allowable battery operation. With just 0.3 Volts more on the battery, the current rises rapidly. At this excessive current, the positive plate will shortly be oxidized and turned to lead dioxide. Ultimately the continuity between the positive terminal and the active material in the positive plate will be lost, and total failure will occur.

Conventional flooded units should be floated toward the high end of the range, while sealed units should generally be floated closer to the lower range. Note that the voltage doesn't continue to decrease as temperature rises above $50°F(10°C)$. This is because the self-discharge current rises as the temperature increases, and to be guaranteed to make up for the self-discharge current, an over voltage is necessary.

Excessively high voltages for long periods will generate internal pressures which will eventually be vented through the safety valves. Repeated venting will lead to drying out of the electrolyte and eventual failure. To prevent excessive charging it is recommended that charging be accomplished with precisely adjusted regulators for the temperature of the battery.

Whereas the electrolyte of a conventional flooded battery stratifies into heavy and light layers, periodic equalization is necessary. The immobilized electrolyte does not stratify ... no equalization is necessary. In practice, it is found that cell voltages in some sealed batteries tend to equalize during float, rather than diverge as the wet battery does. With no need to equalize the sealed battery, which usually involves manual operation, fully automatic regulators are more applicable.

As in the case of conventional batteries, floating at a constant current is better than floating at a constant voltage. This is not

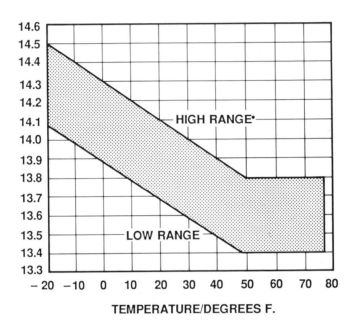

Figure 3.2: Float Voltage versus Temperature

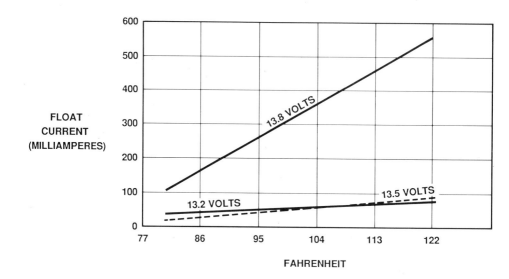

Figure 3.3: Float Current versus Temperature

possible if the battery is also being used as it is being charged. If a battery is to be floated for a long time, a trickle current of 0.04–0.2% of Ah capacity is appropriate. At higher temperature, more trickle current is necessary to maintain the charge. For temperatures below 100 F, a current of 0.04-0.1% is adequate. For a 100 Ah unit, this translates to 40-100 milliamperes. As noted above, some sealed batteries have a low rate of self-discharge and, once fully charged, can be left unattended for many months.

3.7   Cost of Ownership

The true cost of any battery is made up of original invoice price, and the additional expenses which are incurred over the life of the battery. The invoice price of a sealed lead acid battery will be 10–20% more than a top of the line old fashioned battery with equal Ah capacity. If you add gas recombinant caps to the conventional battery, the invoice price may be nearly identical. With charging costs considered, the sealed battery is actually less expensive to own due to its inherently higher charge rate. The other intangible benefits are many, and might be considered as insurance. You won't spill acid in the bilge on a knockdown. You won't have to buy a new battery when you accidently leave your running lights on for a couple of weeks. You won't be constantly adding water. You have the freedom to use some *out of the way* space for your batteries. You won't be spending valuable time equalizing. If you should flood your battery compartment, the battery itself will survive, though it will discharge through salt water.

3.8 An Actual Torture Test

If you do discharge one of the DIN rated units down around the 8 Volts region or lower, you may have to completely discharge the battery before charging. In this regard, the sealed battery acts more like a Ni-Cad battery. Our first experience with this subject area occurred when an instrument with an LED display was left on for about three weeks while we were away. On return the battery had discharged to about 2.5 Volts. Instead of immediately charging the battery fully, we applied a charge for a few hours and then had to leave for another few days before we could finally attend to the battery. When we did return to fully charge the battery, absorption rate was severely low. As the charge progressed, the battery got warm, and although the voltage returned to normal, on removal of the charger, the voltage fell to about 11 Volts in a few hours. We left the charger attached for over 48 hours, hoping that the battery would recover. It didn't.

The battery would sustain a load at about 11 Volts, however, indicating that perhaps the problem lie in a single cell. This was reminiscent of problems we had witnessed with Ni-Cad batteries which can have cell reversals under some overdischarge conditions. We treated the sealed battery to the same cure that we had used for Ni-Cads. First we applied a 10 Amp load on the battery for 24 hours. The load reduced the battery voltage to less than one Volt. Next we put a short circuit on the battery for another 24 hours, using a #12 wire. After that, we put the unit back on the charger with a constant current of about 10% of the batteries Ah rating (10 Amps). Initially, the internal resistance of the battery was high and the charger voltage rose to about 12.5 volts. Shortly, however, the resistance declined and to maintain a 10% current, the charger voltage fell to less than 11 Volts. Battery temperature barely rose above ambient after several hours of charging, indicative of charge acceptance.

We left the charger on constant current, but set the upper voltage limit at 13.8 Volts and left the charger running for a full 72 hours. At the end of the charge, we began a discharge to measure the actual capacity. The unit only had about 60% of its former capacity. Again, we charged the battery at 13.8 Volts for another 72 hours, and measured the capacity again by discharging. Capacity had risen to 85% of rating. After a third 72 hour charge, capacity had recovered to about 95%. Despite the torture applied to the battery, it fully recovered.

You can kill even the best of sealed batteries, however. The best way to do so is to discharge it about 80% and then leave it that way for a few months.

3.9   Useful Life

The *wear out* phenomena is a complex function of the frequency and depth of discharge, the amount and duration of overcharge, and the length of time between a deep discharge and recharge. Time spent at elevated temperature is also a factor in eventual failure. In Figure 3.4 daily capacity loss is plotted against battery temperature. This loss occurs whether the battery is used or unused. It demonstrates quite clearly that batteries should not be hot. While we include this particular figure in the section on sealed batteries, the same phenomena occurs with conventional batteries ... perhaps even worse. The capacity loss shown is permanent, so keep your batteries cool.

3.10   Summary

The sealed lead acid battery is position independent, can be charged fast, requires no equalization, can be left unattended for many months, stands up to high current loads, never needs maintenance, is easily charged and floated, provides a great number of deep discharge cycles and even survives up to 30 days in a completely discharged condition.

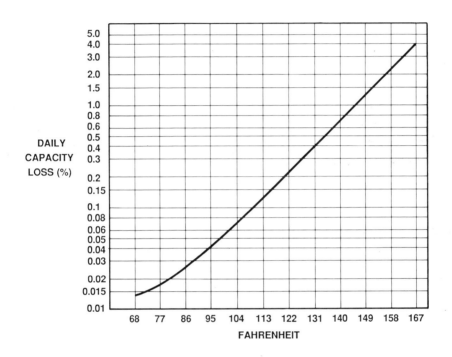

Figure 3.4: Capacity Loss versus Temperature

Besides all these advantages, the cost of ownership may actually be less than that of high quality, old fashioned batteries. Is nostalgia for old fashioned technology worth the inconvenience? For a 1941 Chevrolet pickup, perhaps, but not for a battery built with 1941 methodology. This was demonstrated to us quite effectively when we were pooped by a rogue wave and knocked down in a North Pacific storm. The smell of battery acid in the bilge screamed a simple imperative ... get sealed batteries!

As we pointed out earlier, however, all sealed batteries are not equivalent. As a commodity item, the sealed battery is a newcomer, and as the traditional battery manufacturers begin to develop sealed technology, expect a plethora of sealed units which do not perform even as well as conventional batteries.

Chapter 4

DC Alternators

4.1   General Information

DC alternators are actually three-phase synchronous *AC* generators
with the three outputs connected to silicon rectifiers. The rectifiers,
or diodes, convert the *AC* to *DC* so that it can be stored in batteries.
There are 6 diodes in the typical alternator, 2 per phase. Sometimes
there are three additional diodes called a *diode trio* which are used
to enable the alternator regulator. Figure 4.1 is the schematic of a *P*
Type alternator. The field winding is wound on the rotating center
of the alternator, the rotor. Two carbon brushes conduct electricity
to the rotating winding. One brush is returned to ground, while the
other brush is driven by the regulator. By driving the brush marked
F with a positive voltage, the alternator will charge.

Figure 4.2 shows the *N* Type alternator. The difference between
the N Type and the P Type shown in Figure 4.1 is in the way the
brushes are connected. Note that the *N* Type is excited by grounding
the brush labelled *F*, since the other brush is tied to the positive
output.

Figure 4.3 shows an *N* Type alternator which incorporates three
additional diodes. The output of the additional diodes is used to
power the regulator, thus no enabling switch is necessary. There is
always some residual magnetism to start the alternator charging.

There are two windings of wire in an alternator ... the field wind-

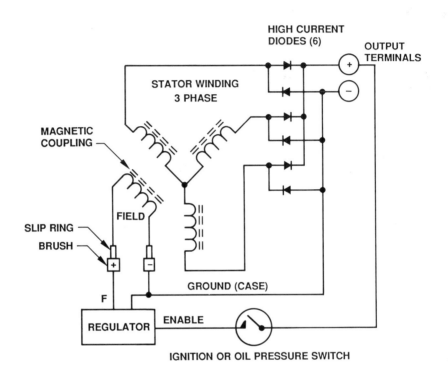

Figure 4.1: *P* Type Alternator Schematic

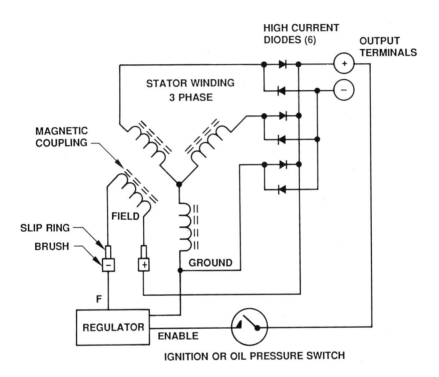

Figure 4.2: *N* Type Alternator Schematic

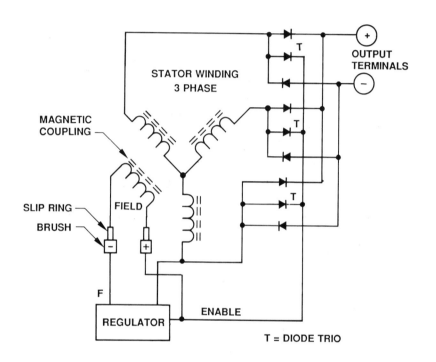

Figure 4.3: Self–Excited N Type Alternator Schematic

Alternator *RPM*	Alternator Amps
2000	30
3000	76
4000	98
5000	112
6000	123
6500	131

Table 4.1: Alternator Output veusus *RPM*

ing and the stator winding. The field winding is more accurately called the control field since controlling the current through the field controls the output of the alternator. In most alternators, the field winding is wound around iron which rotates as the alternator is turned. The iron and winding assembly is called the rotor. The field winding does not need to be rotating ... some very expensive alternators have the field winding interleaved with the stator winding, spinning only the iron rotor.

Alternator output comes from the stator winding. The stator is stationary, and the wires associated with the stator are larger than those of the field since they must carry large amounts of current. The stator winding is actually three windings which are joined at the ends to yield a 3 phase circuit. The way the 3 windings are joined varies, but the effect is the same.

Typically, alternator output current peaks at about 6000–7000 *RPM*. This means that they must be *geared up* by a factor of 3–3.5 for full output on a slow rotating diesel.

Figure 4.1 shows the relationship between *RPM* and alternator output. The output is at a temperature of $200°F$. The alternator is Catalog #1024, from Ample Power Company.

It would be useful to get full output at idle, but don't risk changing the pulley sizes to accomplish this because you may very well destroy the alternator by turning it too fast when you rev up the engine. Designers strive to keep mechanical tolerances within the alternator fairly tight so as to achieve maximum output. If the alternator is rotated too fast any unbalance in the rotor can cause vibration and self-destruction. If you must have more output at low *RPM*, you will have to get a larger frame alternator which has more iron in both the rotor and the stator laminations.

An alternator works on the same magnetic principles as the transformer, but the input frequency is a result of electromagnet rotation rather than *AC* applied to the primary, as in the case of a transformer. The iron in the rotor is energized with a *DC* current. As the rotor spins, alternating *north* and *south* magnetic poles induce current in the stator winding via magnetic flux coupling. Output rating is dependent on the magnetic intensity produced by the field, the amount of magnetizable iron and steel laminations, as well as the *RPM* of the rotor.

As mentioned, it is the field current that is used to control the alternator output. The higher the field current, the higher the output. By varying the current through the field winding, we can control the alternator output. Typically, field current is from 2–6% of the output current. A 100 Amp alternator would require from 2–6 Amps field current. By connecting the field to an electronic regulator, alternator output can be automatically adjusted to meet various load demands. An alternator can be regulated to output a constant voltage, or a constant current. Either mode is accomplished by varying the field current to achieve the expected output. The two modes of regulation are easily combined, so that an alternator may be adjusted for a constant voltage output, subject to current limiting. It only stands to reason that an alternator should be regulated to charge deep cycle batteries properly ... as we know from preceding chapters, multi-step regulation is required, since no single setpoint voltage is correct for both charging, and floating.

4.2 Temperature and Performance

Temperature is the most serious limit imposed on the alternator. Most alternators have no elevated temperature specification. A 75 Amp unit is guaranteed to put out 75 Amps at $77°F(25°C)$. You are running it at your own risk above this temperature, and the minute you start to use the alternator the temperature starts to rise. In the tropics, you're operating the alternator above its specified limit before you begin to rotate it! Note that a cold alternator can produce as much as 135% of its hot rating. Keep this fact in mind when choosing isolator diodes.

The rotor is the very core of the alternator. The field is wound with copper wire. As much as 100 Watts of energy is dissipated in the wire, and as it gets hot, resistance increases, causing a reduction in field current and a corresponding reduction in alternator output. Magnetic material is also less effective as temperatures rise, and can be permanently incapacitated by high temperature. Keeping the rotor cool is a prime concern in a well designed alternator. Much of the heat from the rotor is conducted out of the alternator via the bearings which support the spinning electromagnet (rotor). A quality alternator has high temperature rated bearings.

It's a fact that alternator reliability[1] halves for every $18°F(10°C)$ rise in temperature. The diodes which are a necessary element of the DC alternator, are sensitive semiconductor devices, and prone to failure due to excessive heat. While it makes economic sense to package the diodes with the automotive alternator, it does mean that the diodes are operating under constant stress.

When fast charging batteries, you can expect a diode to fail sooner or later. Actually, most alternators diodes are easily replaced, and they are relatively inexpensive. The real damage of diode failure is usually the wiring to the batteries, and the batteries themselves. A diode usually fails in the short circuit mode, shorting the batteries through the alternator winding to ground. That will burn insulation from wiring, and may likely buckle the plates in a thick plate deep cycle battery. If you are charging both batteries at once, you may find yourself with complete electrical system failure and a fire on your boat, caused by failure of a single diode.

You should take two precautions against such event. The first is simplest and most cost-effective. Place a selector switch on the output of the alternator such that both batteries can be charged separately. Individual charging is a good practice for other reasons, in any case.

---

[1]If you are really interested in reliability projections, refer to Mil Standard HDBK-217D.

Secondly, we recommend a *Fail Safe Diode*[2] (*FSD*) on the output of the alternator. Such a diode should be rated for more than the alternator, and mounted on a heatsink of suitable surface area. A 160 Amp alternator should have a diode rated at a continuous 200 Amps. Diodes can withstand a very high temperature, but for best results, temperature should be kept below boiling. Schottky diodes have a lower voltage drop than do silicon diodes. They will thus waste less energy, and consequently, run much cooler than a conventional diode. Only a Schottky diode is suitable as a Fail Safe Diode.

The Fail Safe Diode will prevent the alternator from shorting out the battery, should a diode in the alternator fail. A Schottky isolator performs like a *FSD*, and if sufficiently rated, will protect against alternator failure. Either an *FSD* or an isolator must get free air movement to stay cool. Isolators which are covered with epoxy will not run as cool as those which are left exposed to the air.

When replacing an alternator, always replace it with a *kkk* rated unit. The *kkk* is a specification of SAE[3] which provides elevated temperature specifications for emergency vehicle alternators. Even with a better alternator though, we still recommend the charge selection switch, and/or *FSD* to limit damage when an alternator diode fails.

4.3   Field Current Control

The simplest of the field current controllers is nothing more complex than a variable resistor. In use, you pay for simplicity by eternal vigilance since you must continually monitor the charging current and battery conditions and adjust the resistor accordingly. Should you forget to reduce alternator output, and continue to pump your batteries with heavy current past the gas point, batteries and alternator may melt into the bilge.

The next level of complexity involves a variable resistor coupled to a semiconductor. The semiconductor is in turn coupled to a voltage sensing *tunnel diode*. The tunnel diode shuts off the semiconductor once the gas point is reached. This system forgives forgetfulness, but

---

[2]Ample Power Company, Catalog #1029.
[3]Society of Automotive Engineers

falls short of performance charging, which provides many features beyond the simple control of alternator field current. For instance, it does not provide for the absorption cycle so necessary on conventional lead acid batteries, since it trips out at the gassing voltage.

Neither the variable resistor nor the resistor/semiconductor field current controllers can provide active regulation. That is, neither senses the voltage of the battery and adjusts alternator output to compensate. The result of this non-active control is that the output of the alternator will *droop* as the battery voltage rises with the charge. You could periodically adjust the controller during charge to maintain the desired current, but again, by constant vigilance.

Because these controllers do not actively regulate, they require that *RPM* remain constant for a constant output current. This is not normally a problem if you are running your engine strictly to charge batteries, but don't try to use simple controllers when you are using the engine to maneuver.

A regulator, on the other hand controls field current and alternator output current in accordance with the charging needs of the battery ... a task that simple controllers cannot undertake. Regulators do not usually drive the field with a steady current proportional to the desired output. It is more typical for the field current to be pulsed on and off, where alternator output is proportional to the average of the pulses. Pulse averaging is accomplished by the inductance[4] of the field winding. We mention this fact, because the rapid on/off switching of field voltage contributes to electrical noise generation. Electrical noise is a significant problem with most alternators to begin with. It is good practice, therefore, to locate alternator field drivers as close to the alternator as possible so that the wires carrying the field current do not act as antennas. Regulators with field switching components mounted in control panels do not meet this design practice and will increase the problems you may already have with radios and direction finders when the alternator is running.

---

[4]Inductance is a property of a coil of wire which impedes a change of current flow, thus applied voltage pulses do not cause current pulses, rather the current is an average value of the applied voltage.

4.4   Alternator Types

Alternators are type cast as either *P* or *N*. A *P* type alternator is one which charges when a Positive excitation voltage is applied to the field. Refer to Figure 4.1. Negative excitation voltage must be applied to the *N* type, shown in Figure 4.2. Note that the voltage applied to the field by the regulator is negative with respect to the other end of the field winding, not truly negative.

Occasionally, a rare *PN* alternator shows up. On the *PN* alternator, both field wires are brought to the outside for external hookup. A *PN* can be wired as either a *P* or an *N* type, thus it is more flexible.

Because there are several types of alternators, some way of determining their type is needed. The *PN* alternator is easy to spot ... it has two field wires going to the regulator. Between the *P* and *N* types, the problem is more difficult. To make matters worse, some alternators have a *diode trio* internally that connects to one side of the field, so that resistance[5] measurement to the case or the positive terminal can be misleading. This is shown in Figure 4.3. All the alternators that we are familiar with, which have a diode trio, are *N* type units.

If you have an alternator with an external regulator, look for a wire marked *F* for Field. Remove the wire which is on the alternator, and with the ohmmeter, measure the resistance between the alternator field terminal and case. Write that reading down. Now measure the resistance between the alternator field and positive output terminal and record the answer. If there is no internal diode trio, one of the readings should be on the order of several Ω. If the reading from field to case is several Ω then the alternator is *P* type. Note in Figure 4.1, that measurement from the F terminal to case will actually measure the resistance of the field winding. If the reading from field to the positive output is several Ω, then you have an *N* type. If you got high readings in both directions, chances are good that the alternator has a diode trio. Sometimes, however, you need to rotate the alternator slightly as you make the measurements so that the brushes make good contact with the slip rings. Without good contact, erroneous resistance readings will result.

---

[5]Resistance is measured in Ohms. The symbol for Ohms is the Greek letter Omega, (Ω), read as Ohm or Ohms.

If you don't have a remote regulator, then the alternator has an internal regulator. The wires going to it should be disconnected. Getting to the field wires is the greatest difficulty. One and sometimes two cases stand in your way.

There are three ways that the alternator field can be connected ... one field wire internally connected to the negative output (*P* type), one field wire internally connected to the positive output (*N* type), or one field wire going to the diode trio.

Once you have identified the field wire which is attached to the regulator, then determining the type of alternator can be done with a jumper wire, in lieu of actual resistance measurement. You will have to run the alternator to do this test, so be careful not to get hands and arms caught in the belt driving the alternator.

With the regulator removed, touch the field wire to ground. If the alternator charges with the field grounded, then it is an *N* type. If not, touch the field terminal to the positive terminal. If you now get charge current, then the alternator is a *P* type. If connection to case or connection to positive do not cause the alternator to charge, then you do not have the field wire, or the alternator is no good.

With the field excited with a jumper as above, the *full fielded* alternator will charge at maximum output. Do not leave the jumper connected for more than 2 seconds.

Re-assembly of the alternator cases can be a puzzle ... how to hold the brushes back so that you can slide the rotor into place. Look for a small hole which is about the size of a toothpick, and aligned so that a toothpick in the hole will keep the brushes out of the way of the shaft during re-assembly. Such a hole will appear in the brush housing and extend out through the case.

4.5 High Voltage Dangers

If an alternator is suddenly disconnected from its load, the output voltage can rise to several hundred volts. As long as the regulator is still functioning, the high voltage excursion will only last for a short time ... but long enough to do severe damage. Not only can all the diodes in the alternator blow out, along with your regulator, but all your electronic equipment may very likely fail as well.

This is the reason we are all cautioned against turning the battery selector switch while the engine is running. The only safe way to do so is to first interrupt the field current. As insurance against someone accidentally turning the selector switch, or the possibility of the switch contacts vibrating open periodically, a transient suppression device or *snubber* should be installed. This device absorbs the high voltage transient until such time as the energy stored in the inductance of the field winding collapses and alternator output returns to normal.

The snubber limits alternator output to about 25 volts, which is much preferred to the several hundred volts which result with no snubber connected to the system. Most electronic equipment can tolerate 25 Volts over voltage, for the very short duration it takes the snubber to operate.

If you use a snubber, however, be sure that the regulator can sense the voltage right at the alternator output terminal. If the regulator is wired to sense battery voltage, such as is proper when isolator diodes are used, then opening the alternator output will result in the regulator losing its feedback. In turn, the regulator will produce more field current in an attempt to boost battery voltage. The net result will be a cooked snubber and eventual destruction of the alternator diodes. A well designed regulator will sense battery voltage, as well as additional sensing directly from the alternator output which operates if the alternator is disconnected from the battery. Such a regulator with a snubber is the only insurance you have if your boat or RV is chartered occasionally.

4.6   Output Characteristics.

Like the battery charger, an alternator does not generate pure *DC*, but instead provides *DC* pulses of current. While the charger pulses at a 60 Hz rate, the alternator pulses at a rate governed by the number of stator windings (3) and the rotational *RPM*. The faster the rotation, the faster is the pulse rate. Because the output is not pure *DC*, the alternator is not suited to floating a battery at less than 13.2 volts.

A 200 Amp alternator produces an average of 200 Amps. Considering that actual conduction may only be 30% of the time, you can see that peak currents must be very high to average 200 Amps. These large peaks are more detrimental to thick plate deep cycle batteries than the cheaper thin plate variety. Under the circumstances, you should probably not use as big an alternator as average current calculations might allow for thick plate batteries. Later in the book we discuss the issue of alternator rating versus battery size and type.

4.7 Electrical Noise Reduction

Current pulses of several hundred Amps through many feet of wire generates significant noise levels. Keeping this noise contained is a major undertaking. Alternator manufacturers make noise suppression even more difficult by using the alternator frame, and consequently your engine mounting bracket, as ground.

We recommend against using engine block and mounting brackets as conductors. In a boat, some electrolysis problems can be traced to the fact that the alternator frame and hence the engine is used as a conductor ... the prop shaft leads directly to the water. A real purist will insulate the alternator from its mounting bracket. While there may be some electrolysis during engine starting, it must be eliminated during the time that the alternator is running. To insulate the alternator, enlarge the mounting holes enough for nylon bushings to be inserted. Use nylon washers under the nuts which are used for mounting. Some alternator manufacturers totally insulate the electrical circuits from the alternator frame ... a policy we wholeheartedly endorse.

Noise can sometimes be cured in low current battery chargers with so-called power filters, but any one at all skilled in engineering would not consider such a solution for the high power current levels of the alternator. The misapplication of a filter designed for other purposes will likely result in a short circuited alternator, and burned up wiring! In particular, avoid the cheap filters sold at automotive parts stores. They may perform in an automobile with a small alternator regulated at a constant voltage, but under heavy long-term charges experienced in performance charging systems, look out. Only install a filter designed specifically for alternator noise suppression, and be sure it is rated for the maximum current that the alternator can produce when cold. Allow at least a 20% margin of safety in the filter's capacity ...use a 200 Amp rated filter for a 165 Amp alternator.

A good quality polypropolene capacitor of 1 micro-farad at 50 Volts can be used to suppress some electrical noise. If you can't find a polypropolene type at a local electronics distributer, ask for a mylar film or metallized mylar film capacitor of the same rating. Place the capacitor across the positive and negative terminals of the alternator.

There are measures which can be taken to reduce, if not eliminate, electrical noise from the alternator. First, use larger wire than what might otherwise be called for. This will keep the end to end voltage drop less. Secondly, put a real ground wire from your system's negative distribution terminal to the alternator frame itself. Your starter is probably also connected to the engine block, so tie the two together with AWG #00 wire, such that no current has to flow through engine chassis or engine mounting blocks to return to the batteries.

These measures probably won't cure all the noise problems. Your next step is to shield the alternator output lead, all the way to the battery. Shielding can prevent radio frequency emissions from the antenna-like output wire. Braided shield wire works best, and can sometimes be found in rolls at industrial surplus outlets. If you can't come up with braided shield, the next best solution is to put a slow wrap of small wire (say AWG #22) around the alternator output lead. The more wire in the wrap, the better the effect. Either the braided shield, or the wrapped shield should be connected to ground on the battery end only, eliminating any possibility of ground loops. As noted above, you should have both a positive and negative wire going to the alternator instead of using engine beds as the negative wire. Twisting these together will reduce electrical noise much the same way that shielding does.

By all means, avoid long wiring runs. In particular, don't route alternator wires to a remote Amp meter, such as in an instrument console. It is better to use short wires from the alternator to the batteries and sense alternator current via a shunt. The shunt should be located close to the alternators and batteries. The voltage across the shunt can be routed to remote meters without generating electrical noise. (The signals across shunts are extremely low level voltages, and routing wires long distances directly from the shunt can incur noise pickup in the wires. This will result in erratic current readings. Noise pickup in shunt wires can be prevented by using a current sensor pre-amplifier which is installed close to the shunt.)

Another source of noise from charging circuits is due to the sequence of connections made at both power and ground terminals. We cover the reasoning behind proper grounding connections in later sections, and won't repeat ourselves here.

Any remaining noise will have to be removed at the input to the electronic equipment which is malfunctioning due to noise. For this application, a capacitor made from ceramic material is cost-effective and generally electrically effective as well. Ask your electronic distributer for a 0.1 micro-farad ceramic capacitor such as a CK05 type, and install the capacitor between the power input to the electronics package. Filters are also available which not only reduce noise, but also provide some energy storage so that engine starting transients don't upset the electronics. For sensitive electronics, an *access diode* with a filter may be necessary. A separate battery for the electronics is sometimes advised. More details about access diodes is presented later in the book.

4.8 Using Isolator Diodes

We are of the opinion that isolator diodes have caused more problems than they have solved. The objective behind their use is to allow charging of two or more batteries from one alternator without connecting the batteries together. Properly installed, this objective can be attained, but rarely are isolators properly installed.

Before we describe the correct way to install isolators so that you can still charge your batteries to the full voltage of the alternator regulator, we feel bound to question the objective itself. Our preference is a single Fail Safe Diode and a charge switch that allows the batteries to be charged independently of one another, or directly in parallel if desired. We think that the charge switch offers more control, and definitely costs less than isolators. (High quality *FSD*s are not inexpensive.)

The fact remains, however, that isolators are convenient. You don't have to remember which battery it is that needs to be charged. Two *FSDs* is a more reliable solution than a single isolator assembly.

Isolator diodes come in two varieties. Conventional silicon diodes and Schottky diodes. The conventional silicon diodes cause a voltage drop of 1–1.5 Volts to exist between the alternator and the battery. The Schottky version reduces this voltage drop to 0.5–0.7 Volts. From our prior discussion of battery charging at a fixed output voltage, it should be clear that losing any voltage between the alternator and the battery will result in totally inadequate charging. To compensate for the voltage lost in the isolators, the regulator must be directly connected to the batteries. With this connection, the regulator will sense battery voltage instead of alternator voltage.

If you are using a regulator with a single sense line, then that line should be connected to the common terminal of your battery selector switch. As we noted earlier, a well designed regulator allows for direct sensing of multiple batteries and also senses the alternator output as protection against an open connection between the alternator and batteries.

In Figure 4.4 we show an alternator connected to two batteries via an isolator consisting of two Fail Safe Diodes. In that figure, we also show another diode which is connected back to the sense input of the regulator. Assuming that all the diodes drop the same amount of voltage, then the alternator output will be high enough to compensate for the loss in the isolators. For conventional silicon diode isolators, use a silicon diode back to the regulator. If a Schottky isolator is used, then the regulator sense diode should also be Schottky. Be sure to use diodes that are capable of carrying the field current of the alternator since the sense point for the regulator may also be the power input to the regulator. If so, a diode of 5–10 Amp rating should be used, and it should be properly fitted to a heatsink.

To make the modifications as shown will require some amount of rewiring of the regulator. Since the conventional regulator is not suited for charging deep cycle batteries efficiently, you should consider leaving the old regulator intact and install an adjustable or multi-step regulator in its place, such as the Automatic 3-Step Deep Cycle

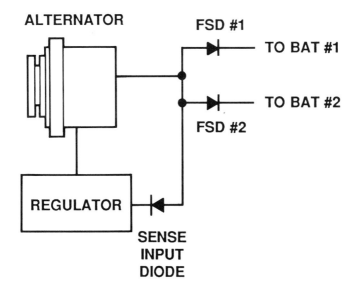

Figure 4.4: Regulator Modifications for Isolator Diodes

Regulator made by Ample Power Company.

Be advised that not all regulators will work properly when connected with the diode shown in Figure 4.4. Without the diode, the regulator senses the average voltage applied to its sense terminal. With the diode, the peak voltage (minus one diode drop) is sensed. The diode prevents discharge back from the sense terminal. If the regulator itself doesn't average the input, then the sensing diode may actually result in a lesser voltage setpoint than without it. You can only find out by trying.

Some manufacturers now build *dual output* alternators. These units effectively incorporate isolator diodes inside the alternator case. (Actually two sets of positive output diodes are used in place of the normal one set.) We find three faults with the dual output alternator ...you now have to measure two outputs and add them to know what the alternator is producing, and the addition of 3 more diodes in an already overheated environment reduces alternator reliability even further. In addition, with two alternator outputs, two filters for noise reduction are required. Because there will be two wires leading to the batteries, instead of one, then noise problems can be that much worse. Our preference for system configurations is summarized in order below.

- Charge Switch with 1 Schottky Fail Safe Diode
- Charge Switch with dual Schottky Fail Safe Diodes
- Charge Switch with Schottky Isolator
- Dual Fail Safe Diodes
- Schottky Isolator
- Dual Output Alternator

4.9   Running Alternators in Parallel

When one won't do, use two. Alternators can be run in parallel, but unless both units are hot rated and can be *run to the rail*, special consideration must be given to how well they share the load. Without using special balancing techniques in the field control current, only identical alternators from the same manufacturer should be operated in parallel. Both alternators should be operated at the same *RPM* if load sharing is to be equal. Even with two identical alternators from the same manufacturer, you should measure the currents individually to verify that the units share the load 50–50. If one alternator is lazy, the other may self-destruct from overload.

With hot rated alternators, it isn't necessary to share the load equally ... without load sharing, one alternator does all the charging up to its current limit. Then the voltage begins to fall, and the other alternator starts to charge. We recommend using high temperature *kkk* rated alternators if you decide to parallel them.

To operate two alternators as one, simply join the two field wires, and also join the output wires. Don't try this with just any regulator though ... the regulator itself is not designed to handle the current of two fields. Make sure to use a regulator rated at the sum of the field current required by both alternators. You should note that only one regulator can be used, not one for each alternator.

Running alternators in parallel is not without special risks. A diode failure in one alternator will short out the other alternator, causing it to fail as well. Rather than directly joining alternator outputs, connect them together via an *FSD*. The diode now protects batteries and wiring from failure, as well as the other alternator. This configuration is shown in Figure 4.5.

Parallel alternators need not always be run together. There will be field current going to both alternators, however, so even the unit not running will get warm. Using one regulator to control the alternators on twin engines is quite appropriate, even if one engine is not running. With only one regulator, there will be no tendency for the two alternators to fight each other which can lead to low frequency voltage excursions.

4.10 Mounting Alternators to Small Engines

In a successful alternate energy system, *AC* power will be used sparingly. In many cases, *AC* can be obtained from inverters for the in-

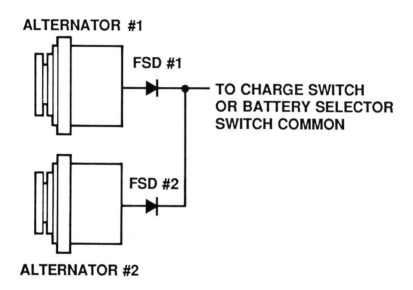

Figure 4.5: Parallel Alternators

frequent times it is used. Charging batteries through an *AC* operated charger is not nearly as cost-effective as charging from an alternator since battery chargers are not usually very efficient and high output chargers are expensive. Thus, you may elect to do without an *AC* generator in favor of a large alternator mounted to a small engine.

Since there is no mass market appeal for small engines with attached alternators, few companies assemble them. On the other hand, there is little magic about such an apparatus, so anyone with mechanical aptitude can build one. Any small engine with a rating of at least 1.5 horsepower (*HP*) is suitable. Beside the engine you will need a pulley, a belt, and mounting brackets.

As noted above, calculate pulley sizes to operate the alternator at 6000–7000 *RPM*. This should occur at the *RPM* which is the best combination of power and efficiency for the engine. Consult the *RPM*/Power/Efficiency curves for the engine you plan to use.

You can lose as much as 1 *HP* just driving a belt under load around two pulleys. This is lost energy. To get the most energy from the engine/alternator you should directly couple them. This requires a precise alignment between their shafts unless a flexible coupling is used. Even then you must pay attention to shaft alignment.

With belt drive, a 1.5 *HP* engine can deliver from 0.5–0.75 *HP* to the alternator. This is just about enough for a 40 Amp alternator. (See Appendix for conversions factors.) Bigger alternators require proportionately bigger engines. In Figure 4.6 we show the horsepower required to drive a 170 Amp alternator[6]. About 9 horsepower is necessary to achieve full output.

You can calculate the *HP* required for a 100 Amp alternator as follows. Assume 100 Amps at a full charge setpoint of 14 Volts.

$$VA = Watts = 1440, \text{ where } V = Voltage, \text{ and } A = Amps$$

There are 746 Watts per *HP* and dividing 1440 by 746 yields 1.93. Doubling this for alternator inefficiency and engine aging yields 4 *HP*. Now add 1 *HP* for the belt. The final result is 5 *HP*. Remember that this is 5 *HP* at the *RPM* chosen. Engine manufacturers usually specify *HP* at the very peak possible. Such a rating is not a good indication

---

[6]Ample Power Company, Catalog #1027.

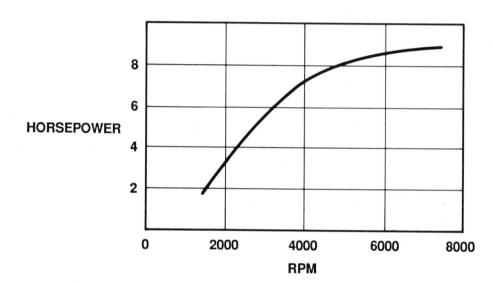

Figure 4.6: Horsepower for 170 Amp Alternator

of usable *HP* in a reliable setting. Derate engine specifications accordingly.

Sizing an engine to an alternator must be done using the peak current that the alternator produces when cold. This can be as much as 135% of the hot rated output. The engine has to be big enough to spin the cold alternator under full load. In this regard, the high output small frame alternators require larger engines than an equivalently rated large frame unit.

Other factors to consider when driving bigger alternators is the number of belts required, and the permissible side load on the engine shaft. If you want to get long life on the system, use two belts. Be sure to check the engine specifications regarding the amount of side load permitted on the engine shaft. Excess side load on unsuitable bearings will result in premature engine failure.

For best performance, use an engine with a governor. Without a govenor, the engine will race when the regulator starts to reduce the charge current, since less energy is required to turn the alternator. There are many good engines to choose from ... consult your local industrial engine dealer.

4.11   Regulation of Alternators

As we noted earlier, an alternator should be regulated to charge deep cycle batteries properly. The standard alternator regulator is a compromise even for automotive batteries. For deep cycle batteries it is a disaster.

The standard regulator is designed to operate at a constant output voltage. Generally, the regulator is temperature compensated, but not for the temperature of the batteries, but rather the temperature of the alternator. Thus as the alternator gets hot, the output voltage is reduced. The net result is less charge being delivered to the batteries. Regulator temperature compensation is just a protection mechanism for the alternator.

The output voltage of a standard regulator is generally set low, but some *high output* alternators achieve their high output by using a regulator set as high as 14.7 volts. While this will charge your batteries fast, it will also cook them fast if you run the engine long. It isn't surprising that so many boaters and RVers go through batteries every season. With a typical regulator at a low constant voltage, batteries are never fully charged. With a high constant voltage, batteries are overcharged any time the engine is run for several hours.

Proper charging requires at least the three steps of bulk charge, absorption and float. Furthermore, the voltage values of each step must be temperature compensated for battery temperature, not alternator temperature. Unless you have a hot rated alternator, some sort of alternator current limiting may be necessary.

A manually adjustable regulator which allows you to set the exact voltage and current is best suited to the job of proper charging. Use of such a regulator requires accurate measurement of both voltage and current, and battery temperature must be accounted for. With the manual regulator, what you set is what you get. An example of this type of regulator is the Multi-Source Regulator sold by Ample Power Company.

Unlike a simple field controller, no constant attention is required for a manual regulator. When you first start charging, set an approriate high voltage (14.4 Volts). Later, as the current through the battery drops to 5–10% of the battery's Ah rating, reduce the voltage to a lower charge voltage (13.8). If you are going to run long hours, reduce the voltage to a proper float voltage when the battery current has fallen to about 1% of the batteries Ah rating. At all times, of course, limit alternator current to a safe level using the Amps control. A manually controlled regulator also allows the batteries to be equalized from the alternator if the regulator can sense battery current and keep it at a constant value approximately 5% of the battery's Ah rating.

A manual regulator cannot be operated by an unskilled person, and many persons do not find the time to gain proficiency. In a charter situation a manul regulator is out of the question.

Automatic regulators which do not provide an absorption cycle but *do* allow bulk charge and float are to be preferred over the standard regulator. Proper float voltages are discussed in Chapters 2 and 3. For best performance where no user interaction is required, a regulator must provide for bulk charge, *absorption* and float. The absorption and float voltages should be separately adjustable at the time of installation, so that system needs and battery types can be accounted for. Once set, the regulator should sense both battery voltage and battery temperature and adjust its output accordingly. A regulator of this type is the only one to meet the needs of an unskilled operator such as a charter client.

An example of an automatic deep cycle regulator is the Automatic 3-Step Deep Cycle Regulator made by Ample Power Company. An alternator connected to both a manual regulator, and an automatic regulator provides the best of both worlds. For day to day charging, the automatic regulator does the job. From time to time, the manual regulator can be used to equalize the batteries, or provide charging under special circumstances. Such a special case arises when it is necessary to relieve the engine of all loads except those absolutely required. In this case, the manual regulator allows charging to be adjusted to a bare minimum.

4.12 Troubleshooting

Failures within an electrical system are most often traceable to mechanical failures. The alternator is no exception. The first and easiest thing to check is the drive belt itself. It must be tight enough on the pulleys so that it doesn't slip, but not so tight that excessive sideload is placed on the alternator bearings. For best operation, drive belts should make at least a 25% wrap around a pulley. If you have less than a 50% wrap, expect belts to wear faster and give more trouble. You should be able to depress a properly tensioned belt about 0.5 inches with a force of about 10 pounds. High output alternators will require more belt tension, and perhaps the use of belt dressing which helps prevent slippage.

After belt tension, check the wire attachments. Any connections made via studs and nuts are suspect, but not as much so as connections via push tabs. Push tabs are reliable if properly sized and not removed more than several times. Intermittent problems in particular are often traceable to loose push tabs.

Regulators are known to fail all too frequently. This is particlarly true of integral regulators which run at the same hot temperature of the alternator. A regulator must get power before it can operate. Measure the voltage at its input ... a faulty ignition wire or switch is often to blame. With batteries disconnected, measure the resistance of the remaining wires which connect the regulator to the alternator. Are they all connected with a low resistance less than an Ω?

Once you are convinced that wires are connected properly, then the regulator itself is suspect. It is easier to test the alternator itself though. You can do that by removing the field wire from the regulator and connecting it to either alternator output or alternator ground (P type, N type respectively). If you don't get charging, then it is likely that the alternator itself is defective. If you do get charge current with the *full fielded* alternator, then the regulator is shot, or it's sense line is open circuited.

Brushes in the alternator don't usually fail per se, they just wear down until contact pressure is insufficient for a good connection to the slip ring. As noted earlier, an ohmmeter can be used to measure field winding resistance. Measuring the field winding resistance through the brushes, expect a resistance of less than 5 Ω. Brushes may sometimes hang up in the brush housing. They should be free to move, pressed against the slip ring by spring pressure.

Under emergency conditions, you can make do without a regulator. Use a switch to control the field. Remember, however that the alternator is producing full output when the switch is on. Don't leave the switch connected long enough for your batteries to be overcharged or alternator overheated. Ten minutes on followed by ten minutes off is probably all right for the alternator. Keep an eye on battery voltage, and when 14.4 Volts is reached, stop charging. The switch should be rated for at least 5 Amps, and 125 *VAC*. Whenever the switch is opened, the field voltage will tend to arc across the switch contacts, and will short a switch not rated as above.

Figure 4.7 shows how a switch and a diode may be used to control the field winding of an alternator in an emergency. With the switch on, the alternator will charge at its maximum rate. Engine *RPM* can be adjusted to limit alternator current to a safe value.

Switch arcing can, and should be prevented by connecting a suitable diode in parallel with the field winding. The diode polarity has

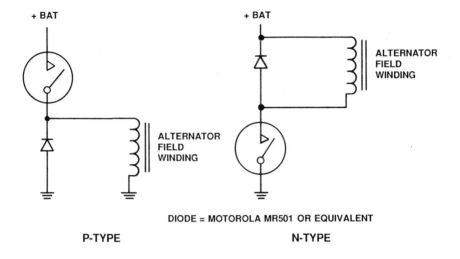

Figure 4.7: Emergency Switch for Alternator

Description	Quantity
Positive Diode Set (3 diodes in set)	1
Negative Diode Set (3 diodes in set)	1
Diode Trio (if used)	1
Brushes	2
Front Bearing	1
Rear Bearing	1
Regulator Assembly	1
Drive Belts	2
Stator Winding and Lamination Assembly	1
Emergency Toggle Switch/Diode	1

Table 4.2: Alternator Spare Parts

to be such that it is normally reverse biased. When the switch is opened, the field voltage reverses polarity, and the diode momentarily conducts until the energy in the field inductance is dissipated.

We have noted several times that diodes in alternators do fail. It is also possible for either the field winding or the stator winding to short out to the case. These failures can be detected with your Ohmmeter. An alternator which has been seriously overheated may have suffered permanent damage to the magnetic material in its rotor or stator laminations. Nothing but replacement can fix this. Often, the bearings are the first parts to fail in an alternator due to excessive side load and extremely high operating temperatures.

4.13   Spare Parts

If you are mechanically inclined and can use a Volt/Ohmmeter, then rebuilding an alternator will be routine. Tabulated in Figure 4.2 are the recommended spare parts for the serious rebuilder.

Everything except the bearings are easily unbolted and replaced. Alternator cases are made by casting aluminum into a mold. The grade of aluminum is usually of low quality and can easily be fractured or otherwise gouged. It is best to avoid bearing pullers and other methods which use undue force to remove bearings. Instead, use the force of gravity and heat to remove them. Remove all diodes and the stator windings from the cases. Now turn them so that the bearings can drop out of the case, and slowly apply heat to the aluminum case. Do not heat the bearing ... the idea is to heat the case. (Use a butane torch, but remember that aluminum melts at about 1200 F.) Aluminum expands under heat much faster than does the steel in the bearings, and the coefficient of expansion per degree is also greater. As you are heating, start tapping the case gently with a rubber mallot. With any luck at all, the bearings will drop out of the case as the case expands.

Installing new bearings can also be done without force. Several hours before you are ready to remove the old bearings, put the new bearings in the coldest place possible, for instance a freezer. Now, when the old bearings fall out of the very hot case, simply drop in the cold new bearing. Since the bearing is cold, and the case is hot, the bearing will fall into place with a satisfying thunk. The aluminum will quickly shrink around the cold bearing tightly.

For most people, carrying a spare alternator is preferable to carrying spare parts. It is much easier to bolt up a new alternator at sea than to rebuild one. The spare alternator doesn't need to be the same high output unit you will normally use, if cost is prohibitive, but it should attach with exactly the same bracketry so that replacement is fast and easy. If you are presently retrofitting to a high output unit, save the old alternator with all of its attachment hardware and carry it as a spare.

A spare regulator is also good to have. If you can't afford two of a proper multi-step regulator, keep your old fixed voltage regulator for an emergency.

4.14 Summary

Your alternator will most likely be your biggest source of electrical energy. It produces pulsating DC current and thus can charge batteries directly. The alternator must be regulated properly to charge deep cycle batteries ... a 3-Step Regulator is required. Alternator diodes run hot and failures do occur. Protect your electrical system with a Fail Safe Diode. Use a snubber to protect against over voltage transients which result if the running alternator is disconnected from the batteries. Either carry a spare alternator, or spare parts to repair a faulty one.

Chapter 5

Battery Chargers

5.1   General Information

Technically speaking, alternators, solar panels, wind generators, and water generators are battery chargers, but it is generally understood that battery chargers are those things which plug into the *AC* power outlet and charge batteries.

Battery chargers come in all sizes and shapes ...their physical diversity only matched by the usually mis-leading claims of the manufacturers. Perhaps you remember the pitched battles which were fought by stereo manufacturers over the subject of watts per channel. Eventually, enough public pressure was applied so that after all the outrageous advertising claims were made about output power, the buyer could read a common truth usually found in the smallest print ...*RMS* watts per channel.

Since the buying public for marine battery chargers is small, it is too much to hope that charger manufacturers will someday be required to put meaningful specifications on their units. It is up to you the buyer to sort through the claims and determine if the chargers are applicable to proper charging of deep cycle batteries.

5.2   Battery Charger Specifications

Battery chargers are usually specified in Amperes ...i. e. a 10 amp unit. The questions left unanswered are these ...at what voltage does the charger produce 10 Amps, and at what input line voltage? One charger we know about claims to regulate as low as 95 Volts on the input. This is a common specification for high quality power supplies such as those used for computer equipment, and it means that the output remains within tightly specified limits until the applied *AC* falls below 95 Volts. As our tests on the battery charger confirmed, however, the output current was a direct function of input voltage, falling to about zero when the *AC* input had fallen to 95 Volts. That's called regulation? We call that obfuscation, which in this case is a euphemism for the word *lie*.

We know another battery charger rated at 70 Amps. And it outputs those 70 Amps as stated until the battery voltage has risen all the way to 11.5 volts! **Eleven and a half Volts?** Most batteries we know can be run flat dead and recover, with no charging, back to 11.5 Volts. Proper use of batteries dictates that they be charged long before they reach 11.5 Volts, and, all the way back to 14.4 Volts. Who is this manufacturer trying to kid? The claims must be effective ...the manufacturer does well in the charger market.

To actually specify is to provide sufficient details so that no misinterpretations can be made to a person skilled in the meaning of the terms used. Misuse of terms, and lack of detail may be good marketing ploys, but a reputable manufacturer does not stoop to use them.

First and foremost, a battery charger that carries an output amperage rating without an accompanying output voltage and input voltage specification is not made by a manufacturer interested in solving your charger needs. A proper specification might be stated as follows ...10 Amps *average* output at 13.8 *VDC* with 115 *VAC* input. We use 13.8 volts here because it is the voltage most battery manufacturers specify as a combination charge/absorb/float value. If, in fact you wish to charge faster, an additional specification would be helpful, for instance, what output current will the charger provide at 14.4 volts? Note also that we specified average current output. Just as the stereo makers like to quote *peak* power figures, we can imagine a specifications war between manufacturers over peak Amps.

In addition to these basic specs, we would also expect to see some details about behavior at low *AC* line voltage such as that often encountered at the end of the dock. Has the manufacturer really designed in enough steel laminations in the power transformer to provide full rated output at 105 *VAC* input, or must you be next door to the transfer station to get full rated output.

Finally, we'd like to see real specs regarding temperature. Is the rated output applicable to the 120 degrees found inside close quarters in the tropics, or is the rated output only available at ambient freezing? How stable is the regulation setpoint with both time and temperature? Will the setpoint stay at the value it was set to, or will you have to re-adjust it periodically?

As we stated above, it is up to you to sort through the claims. If you find the salesman reluctant to supply meaningful specifications, or just as bad, ignorant of the same, find another place to shop. Only wise consumers can buy high quality and functional products. The plethora of barely functional battery chargers belies the intelligence of the consumer.

5.3 Battery Charger Types

Battery chargers can be classified by the type of regulation *topology* used. We use the term topology to mean the basic circuits which are employed. Under this method of classifying chargers we identify 4 distinct types:

- no regulation

- ferro-resonant regulation

- phase controlled regulation

- high frequency *switcher* regulation

Each of these chargers has relative merits and faults.

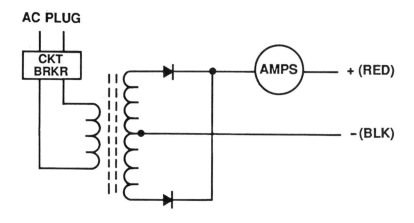

Figure 5.1: Unregulated Battery Charger

5.3.1 No Regulation

In Figure 5.1 we show the schematic of an inexpensive charger often
used as an occasional overnight charger for automotive batteries. As
shown, the charger is little more than a transformer and a couple
of diodes. The transformer is usually wound with as few sall wires
as possible, which provides enough series resistance to limit output
current long enough to allow the circuit breaker to protect the diodes.
The circuit breaker may periodically open up when charging deeply
discharged batteries.

The output voltage is directly proportional to input AC. That
ratio is governed by the turns ratio between primary and secondary
windings. Our experience with chargers made by Schauer[1] indicates
rated output at 115 VAC input and 13.8 volts output. As the battery
reaches full charge, output voltage continues to rise. It will reach 16+
Volts if left on the battery long enough ... boiling all the water out of
the battery if not disconnected. Schauer battery chargers are sold in
discount automotive parts stores at very inexpensive prices.

---

[1]Schauer Manufacturing Corporation, Cincinnati, Ohio

These *cheapie* chargers have features not found in many marine chargers, however. First they are UL[2] approved, assuring the user against the use of dangerous components and manufacturing practices. Secondly, they have an ammeter, however inaccurate it may be. Because the charger has few components, reliability is usually very high. The Schauer charger also includes an isolation transformer so that electrolysis is not a problem. Last but not least, the cheapies include an overload circuit breaker. At rock bottom price, we could only hope they were regulated too.

Posing as a customer in the market for a charger, we have asked several sales persons of battery charger manufacturers why we should buy one of their units rather than an unregulated charger. One of the consistent answers was, "the inexpensive chargers cause electrolysis". That reply is enough to scare any boater and is undoubtably effective. It is also untrue. Any UL approved battery charger must have an isolation transformer[3]. Given our druthers, we'd take a UL approved transformer any day over one that isn't. To satisfy yourself that the charger is isolated, simply measure the resistance between the output leads and the *AC* leads. Be sure to test all wires to all wires. You should read open circuit between any output wire and any input wire. Reverse meter leads for each measurement to make sure that a diode in not blocking the reading. To pass the UL test, a high voltage must be applied between input and output, not just a resistance reading as above.

Finally, an unregulated charger has the highest efficiency rating. This fact makes them ideally suited for charging from a small *AC* generator. Whereas a regulated charger may dissipate 45% of its input power in the charger itself as heat, the unregulated unit will deliver 85-90% of its input power to the batteries, with only a small heat loss. Since you are running the generator, it is assumed that you wish to charge the batteries as fast as possible. No regulation is required under these circumstances. The running generator is reminder enough that charging in in process, and the short length of time that you might overcharge under the circumstances is not detrimental.

---

[2]Underwriters Laboratories Inc.
[3]UL specification 1236.

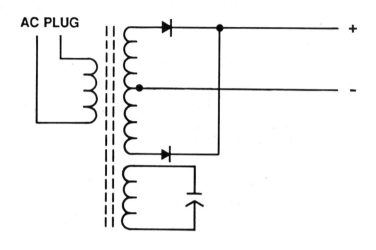

Figure 5.2: Ferro-resonant Battery Charger

5.3.2   Ferro-resonant Chargers

In Figure 5.2 we show how a ferro-resonant charger is made. The transformer is loaded with a capacitor, the value of which is selected to cause transformer saturation at the operating frequency. The capacitor and the extra winding *resonate* at the input AC frequency. This causes the transformer to saturate and therefore to regulate the output voltage. The transformer must be designed for the AC line frequency, thus a 60 Hz unit will not operate properly on 50 Hz.

Saturation is the term used to describe the condition where an increase in input current causes no further increase in output current. A special transformer with a magnetic *shunt* is used to obtain a predictable saturation point. The steel laminations in the transformer cannot be further magnetized, once the saturation point is reached.

As the AC line voltage switches from one polarity to the other, the transformer switches from *positive* saturation to *negative* saturation. Since saturation is a predictable state, governed by the windings, the capacitor value, and the magnetic shunt, a form of regulation is achieved. The regulation setpoint is fixed by the magnetic saturation point and is not adjustable.

If we were allowed but a single word to describe ferro-resonant chargers, it would be *inefficient*. Besides being an energy hog, which has significant meaning if you are trying to use a small *AC* generator, the ferro-resonant charger is characterized by poor regulation, both from an input *AC* point of view, and a load current viewpoint. Unlike a precisely regulated charger, the ferro-resonant charger does not reach its regulation voltage at rated current, and thereafter maintain a constant voltage. Generally the output current starts falling long before the regulation voltage is reached. Recall the 70 Amp charger we mentioned earlier? It barely produced 5 Amps at 13.8 Volts. We should not be too critical of ferro-resonant chargers ... they do make good closet warmers. There may be something else good to be said about ferro-resonant chargers, but we don't know what it is.

5.3.3 Phase Controlled Chargers

The basic circuit for a *phase controlled SCR* charger is shown in Figure 5.3. Because the phase controlled charger uses more sophisticated electronic circuits than the ferro-resonant charger, much better regulation can be achieved, over a wider range of input voltages and output currents. (Yes, it costs more.)

Figure 5.3 is a simplified diagram for a phase controlled charger. The *SCR* is triggered into conduction at the proper phase of the input *AC* frequency. The proper phase is determined by additional regulation circuits which vary the firing of the *SCR* to achieve a constant output voltage. The *SCR* cannot be triggered off, but each second, the *AC* line voltage goes through zero 120 times. The *SCR* turns off each time, and then waits for the phased firing signal.

The *SCR* is an electronic device which can be turned on, but not off. The *SCR*, once turned on, will only stop conducting when external circumstances won't allow current to flow anyway. The term phase control is used to describe the way regulation is accomplished with the *SCR* as prime control element. As shown in Figure 5.3, full wave rectification is performed ahead of the *SCR*. For a portion of each 60 *Hz* input cycle, the *SCR* is conducting. It is *fired* into the on state

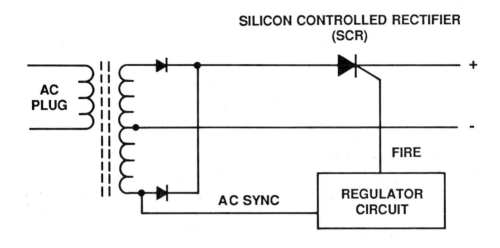

Figure 5.3: Phase controlled *SCR* Charger

at a specific phase of the input *AC* frequency. The phased timing of *SCR* firing governs the average output from the charger. It is up to the regulation circuitry to adjust the proper firing phase to the *SCR*.

The phase controlled *SCR* charger is better regulated than the ferro-resonant unit, and generally more efficient. You do get something for your money. They are also electrically noisy. Noise is generated by the sudden firing of the *SCR* into its conduction state. The rapid rise in current through the charger wiring generates radio frequency emissions, as well as 120 *Hz* noise. As you probably know, *AM* radio bands are most severely affected by *SCR* chargers. The sudden firing of the *SCR* also causes the steel laminations in the charger to vibrate ... hence the audible hum.

The *SCR* is not a particularly efficient device. Because it may drop as much as 2.5 volts when conducting, it can seriously limit charger efficiency (2.5 Volts out of 13.8 Volts). More expensive *SCR* devices have less voltage drop, but their greater price usually limits their use to non-consumer applications.

Phase control can also be accomplished with a relative newcomer to the power handling arena, the metal oxide semiconducting field effect transistor (*MOSFET*). The *MOSFET* is generally not as *husky* as an *SCR* (peak to average current capacity), but they exhibit extremely low voltage drop when conducting, and therefore, if properly applied, can yield significant efficiency improvements over the *SCR*. Unlike the *SCR*, the *MOSFET* can be turned off while conducting.

5.3.4 Switchers

Switching power supplies are the technology of choice in the fast paced computer business. Here, the advantages of high power handling in physically small units can mean the difference between a winner and an *also ran*.

The switcher gains its size/weight/power advantage and its name from the technology used. It all comes down to the transformer. To operate efficiently (95% power transfer) at 60 *Hz*, a lot of bulky and heavy steel laminations are required. Suppose we can increase the operating frequency from 60 *Hz* to say 600, or 6000? Would you believe 600,000? Switching frequencies of 2,000,000 *Hz* are possible, but most current applications are in the range of 25,000 to 250,000 *Hz*. At these frequencies, exotic magnetic material is necessary, as well as the *MOSFET* described earlier, but the reduction in size and weight is dramatic. A 40 pound transformer shrinks into ounces by switching the input at 250,000 *Hz* instead of 60 *Hz*.

Switching technology may not captivate the battery charger market soon, however. Design of high frequency units is not quite yet a cookbook design exercise. The special expertise required to design a switching charger is probably fully employed in higher volume and more profitable ventures than the marine or RV battery charger market. The standard switcher circuit also requires a minimum current load, generally about 20% of full rated load. Obviously this is not compatible with the float necessity of a battery, where load current is close to zero.

There is a switching battery charger on the market today. It gets around the minimum load current requirement by cycling the charger on and off. This kind of operation is not at all compatible with floating a battery at a constant, pure *DC* voltage, and unless you are very concerned about weight on your boat, we recommend you consider other options.

Switchers have a problem other than minimum load current. Because they operate at frequencies close to the commercial radio bands, they can generate electrical noise which is hazardous to essential communications. Cases which house the electronics have to be specially designed to prevent electromagnetic interference from escaping. Even small holes in the case can permit emissions.

5.4   Regulation Requirements

Phase controlled, and other regulated chargers are adaptable to either a constant voltage or constant current output. A Ni-Cad (nickel cadmium) battery should be charged by a constant current, while a lead acid battery is best charged by a constant voltage. A constant current charge is appropriate for a lead acid battery if the regulation is switched to constant voltage for the absoption cycle, and, at completion of absorption, to a constant float voltage.

Because of the different modes of charging, and the different regulation setpoints required at different charge cycles, it is easy to fall into the automatic, multi-cycle battery charger syndrome. We want to believe that this charger or that charger properly cares for our batteries, with minimum, or no intervention on our part. If it were only that easy.

Even the meaning of *automatic* is obscured by some charger manufacturers. Some use the word to mean regulation. Others imply that a change of state occurs in the regulation circuitry. Determining just what is automatic is a problem. Then the question arises, is it appropriate?

There are chargers which offer multi-state charging ability. Some provide a constant current up to a given battery voltage, and then switch to constant voltage regulation. This type of charger must be balanced carefully with the capacity of the batteries being charged. If the charger is too big, it will drive the batteries up to the voltage trip point long before they reach a full charge. In this case the absorption cycle is missed completely, and the batteries are a long way from a full charge.

On the other hand, if the batteries are too big, the charger may never get them up to the voltage trip point. It may drive them high enough to boil all the water from the batteries, but not high enough to automatically trip to a float voltage. From direct experience, we know that gas recombinant caps can melt if left on a charger that hangs up above the gassing voltage.

As we pointed out earlier, batteries from one manufacturer to the next have different optimum charge programs. Even the way a battery is discharged has a bearing on the way it should be recharged. Under the circumstances, the only real regulation requirement is adjustability, and an operator with adequate system instrumentation and the know how to use it. There is no merit in doing something automatically wrong . . . find a battery charger that can be adjusted and learn how to operate it.

5.5 Choosing a Battery Charger

Some of the fanciest looking battery chargers are the worst kind to buy, unless your intent is merely to impress your friends. Before you can really decide on the right battery charger, you must answer some questions.

Is the charger to be used solely or mostly while you are tied up to the dock, or do you expect to use the charger offshore from a generator? Do you expect the charger to put a fast charge on the batteries?

If you only use the charger at the dock, then the most important aspect of the charger will be the output voltage and how well regulated it is. Under these circumstances, you are not using the charger for actual battery charging, but rather to maintain the batteries without boiling all the water out of them. For floating of batteries, the output voltage should be adjustable in the range of 13.2–13.8 Volts. This range of control will be adequate to compensate for temperature and still do moderate charging when necessary.

A battery charger that is to be used for charging offshore must have a wider range of control. If you are using a small generator to run the charger, then efficiency is also very important. You want to deliver all the available power from the generator to the batteries, not waste it as heat inside the charger.

Choosing a battery charger is complicated by the fact that manufacturers do not rate them *fairly*. We use the word fairly instead of honestly because one can state a meaningless specification honestly, however unfair it might be to an unsuspecting consumer. For a charger to be effective, it must deliver Amps at 13.8–14.4 Volts, not 11.5 Volts.

Actively regulated chargers are much preferred to ferro-resonant types. They at least will hold a more constant voltage as the *AC* line varies. Unfortunately, a constant voltage is not the best way to charge. If the voltage is high enough to charge, then it is too high for long battery life.

A battery charger designed for both float service at the dock and fast charges offshore is hard to come by. For float service, a voltage range of 13.2–13.8 is required. For fast charges offshore, we want to run the charger at maximum, delivering as much of the input power as possible to the batteries.

The most efficient battery charger made is the unregulated type. As long as the charger has an isolation transformer it will not cause electrolysis. A small 650 Watt portable generator can be used to drive a 40 Amp unregulated charger. When you are offshore and want to make maximum use of a generator, then an unregulated charger is very appropriate.

A small adjustable battery charger for use when tied up to the *AC* power line, and a larger unregulated unit for charging from a portable generator make a good combination and can probably be purchased for less than a single *marine* charger which is not regulated properly for either bulk charge or float.

5.6   Summary

Battery chargers come in just about every imaginable size, shape and claimed performance ratings. The claims are part hype, but mostly they consist of misrepresentations. Ferro-resonant types are heavy, inefficient and provide sloppy regulation. *SCR* phase controlled chargers are electrically noisy and have an annoying hum. Unregulated chargers can cook a battery if left on long enough. Two-state chargers that trip from a constant current to a constant voltage may trip too soon or never, depending on the match between the charger and the battery bank. We don't know of a single battery charger which, as sold, is good enough for our batteries. At dockside, we currently use a 10 Amp Schauer unit with an external adjustable regulator from Ample Power Company. Offshore we use a 40 Amp unregulated charger which we designed ourselves. It is nothing more than a custom transformer and two Schottky diodes. The 40 Amp charger runs from a portable 650 Watt *AC* generator.

Chapter 6

Solar Panels

6.1 General Information

During the next 15 minutes, more energy than all the energy consumed by humanity during the year will fall on earth in the form of sunshine. It is only natural to wish that this great source of energy could be harnessed to the machinery that drives our abundant lifestyle.

Photo-voltaic cells would seem to hold this promise. Operating silently, with no moving parts, they generate the electricity we want for our machines. Unfortunately, the cost of materials for a solar panel are still high. Likewise, converting sunlight to electrical energy is an inefficient process, with no real breakthroughs on the horizon. The net result is too little energy for too high a cost to make solar panels the object of their promise. Even so, in the last decade the price per Watt of power has fallen about a factor of ten for large solar arrays. Even the small panels have shown a price reduction recently.

Though expensive, solar panels have a place in small electrical systems that are owner, and therefore wisely, operated. A solar panel can spell the difference between a dead battery, and one that can start an engine.

To get maximum use from a solar panel, requires that it receive as much direct sunlight as possible. That means that the panel should be ninety degrees to the suns rays, and the surface of the panel should be as clean as possible. Avoiding scratches in the surface which may diffuse the light is a good practice, so don't clean the panel with abrasive cleansers.

6.2   Solar Panel Ratings

Solar panels are generally rated in Watts. To find out what kind of Amps are available, you have to divide the Watts by the voltage at which the Watt specification applies. Some manufacturers rate their panels at 15 Volts, while others use 16.5 Volts. Those specified at a higher voltage allow for voltage drop across an isolating diode. Assume you have a panel rated at 15 Volts and 30 Watts. That amounts to 2 Amps of current. If you use a Schottky diode that drops 0.5 Volts, then you can drive your batteries up to 14.5 Volts, delivering 2 Amps.

Small panels are available which are rated at 5 Watts ... ratings go up from there all the way to large arrays of panels which produce many kilo-Watts. The power ratings apply to a bright sunny day with the panel properly focused on the sun.

Of most interest to us are the panels in the 35 to 45 Watt area. These produce enough power to be useful, yet they are small enough to be used and stowed aboard a boat or *RV*. Several of them can be used in parallel to provide as much power as you can afford and carry.

6.3   Solar Panel Technology

A solar cell is basically a diode consisting of a silicon *PN* junction. The silicon atoms are bound together by shared valence electrons. The silicon atom has 14 electrons, 4 of which may share bonds with other atoms. When an incoming photon of light strikes one of the shared electrons, it is dislodged. The freed electron becomes more energetic, and if sufficiently close to the diode junction, may jump across the semiconductor and become part of the solar current flow. If the freed electron doesn't jump across the *PN* junction, then it will become captivated again, and the photon that struck the electron goes for naught. The re-captivated electron does generate heat by jumping around and jostling other electrons in the process.

Bright sunlight carries about 1000 Watts of energy per square meter of exposed surface area. The idea behind a solar panel is to convert as much of this energy as possible to electron flow rather than to heat and other losses such as reflection. The theoretical maximum conversion efficiency of the silicon cell is about 28%. About half of this is achieved in practice.

In the past, commercial solar panels have been constructed of crystalline silicon. Such silicon has the regular lattice structure of a typical crystal. This results in the silicon atoms aligning themselves with one another. Such alignment means that incoming photons of light will not have a great chance of striking a free electron. To get around this alignment problem, solar cells have been made from thick silicon layers. The extra thickness increases the odds that a photon will strike a free electron.

Recently, a silicon structure called amorphous silicon has been used to make solar panels. In amorphous silicon, the atoms do not align themselves in a regular crystal structure. This means that a photon has a better chance of striking a free electron, and so the silicon can be used in much thinner layers. This new type of cell is thus labelled as an amorphous thin film type. Pure silicon is expensive, so any technique to use less of it means less cost to the user.

Reliability issues of thin film amorphous silicon are less established than the crystalline cells. Output may fall off with age faster than is typical with the thicker crystalline types. In particular, life expectancy of the new flexible solar panels which use amorphous silicon, has yet to be determined.

6.4 Solar Panel Output Characteristics

A solar panel is made up of individual cells. Each cell develops a voltage of about 0.45 Volts. A typical panel to charge 12 Volt batteries will have from 32 to 37 cells in series. The more cells in series, the higher the output voltage.

When sufficient sunlight strikes the panel, electricity is produced. When there is no light, there is not only no electricity, but the solar panel actually conducts current from the batteries. For this reason, solar panels must either be disconnected during dark hours, or prevented from discharge by an isolation diode.

While output may be specified as 44 Watts at an output voltage of 15 Volts, direct output of the panel can reach as high as 20 Volts. If a solar panel is to be left permanently attached to the battery, some sort of regulator is necessary to prevent damaging overcharges.

6.5   Solar Panel Hookup

As mentioned, a solar panel must either be disconnected when the sun isn't shining, or the solar panel must be isolated with a diode to prevent battery discharge.

The best type of isolation diode is one made with a *Schottky* metal/silicon junction. The forward conducting voltage drop is less for this type of diode, allowing the solar panel to deliver more energy to the batteries.

Figure 6.1 shows the proper hookup for a solar panel. A diode is necessary to prevent discharge when the sun is not shining. The diode must be rated for the maximum current. Use one diode per panel array, or one diode per panel, as shown by the dashed lines. As shown, output is either connected to the common point on the battery selector switch, or to the common point of a charge selector switch. Do not use diode splitters or *isolators* with the solar panel if you are using our suggestion about resting batteries so that you can determine 50% discharge. In this case, you should always charge the *in-use* battery with the solar panel. Connecting the panel to the common of the battery selector switch (through an isolation diode) accomplishes this best.

If you connect the solar panel to both batteries by using two diodes instead of one, then neither battery will receive the rest it needs for electrolyte stabilization. Without stabilizing rest, you will not be able to determine capacity remaining in the battery. That's ok, if you have big enough solar panels to keep up with demand, but if you rely on supplemental charging from time to time, then you need to know the state of charge on your batteries so that you can schedule charging from an auxiliary source.

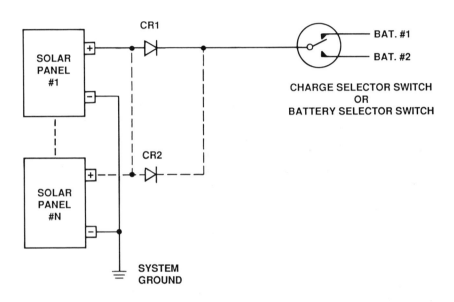

Figure 6.1: Solar Panel Hook–Up

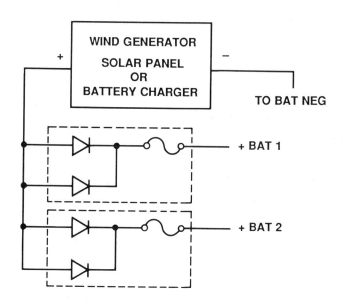

Figure 6.2: Connection to two Batteries

Figure 6.2 shows a solar panel arrangement which allows the panel to charge two isolated batteries. This arrangement is permissible when you do not need to measure the voltage on a rested battery. The Charge/Access Diode shown is made by Ample Power Company.

Note that some solar panels include an isolation diode already connected to the panel. If another *splitter* diode is added in series with the built-in diode, then the effective panel power will be reduced.

6.6   Solar Panel Regulation

Recently, solar panels have been introduced that are *self regulating.* It sounds good, but in fact the solar panels do not regulate at all. What the manufacturers have done is reduce the number of individual cells in the panel so that the output voltage is less. With less output voltage, there is less chance that the panel will overcharge the battery. We suppose that because the panel *self regulates*, a manufacturer can charge more for the *benefit*. Don't go for the self regulation ploy unless you only want to use the panel for float charge. If you want to do serious charging, stick with the higher output panel and be prepared to regulate it if necessary.

Solar panels are generally regulated with *shunt* regulators. The word shunt is not to be confused with the resistive shunt used for current measurements, though the two uses of the word are similiar. A resistive shunt is so-called, because it shunts or passes the majority of current, requiring the measurement circuitry to pass little. A shunt regulator operates to shunt or conduct solar panel current to ground rather than into the batteries. Shunt regulation is effective but the regulator and solar panel must dissipate the energy which would otherwise go into the battery. Shunt regulators therefore tend to operate hot, with lesser reliability the consequence.

Shunt regulators can work in the linear mode, conducting just enough current to ground to keep the battery voltage at the desired setpoint, or the shunt regulator may work in the switched mode. In this latter mode, the solar panel is shorted for some period, followed by a period when all the solar panel output is coupled to the battery. By varying the duty cycle of the shorting period, an average current can be delivered to the battery which keeps the voltage at the desired setpoint.

A solar panel can also be regulated by a series switch which is turned on and off to keep the battery at the desired setpoint. The switch, of course, must not be mechanical because it would not last long under a high frequency duty cycle. Solid state switches capable of interrupting the solar panel current must not, when conducting, drop much of the available voltage. This is a tough order to fill, which explains why shunt regulators are generally used.

Figure 6.3 shows how a solar panel can be regulated with a series switchmode regulator, and how the panel can be instrumented. The 3-Step Regulator is made by Ample Power Company. The regulator turns on and off at a rapid rate, where the output voltage is determined by the on/off ratio. Note that the output goes to the selected

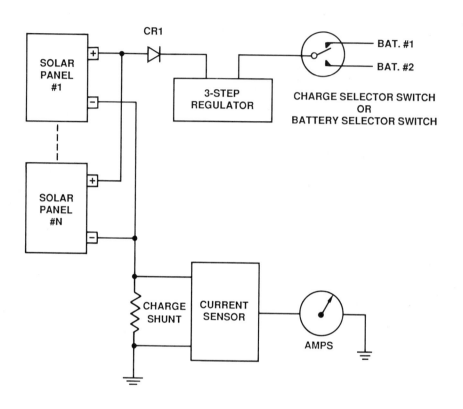

Figure 6.3: Solar Panel Regulation and Instrumentation

battery.

As with battery chargers, and alternators, the key to optimum preformance is regulator adjustability necessitated by the battery.

6.7   Using Solar Cells to Float Batteries

From Chapter 2 we learned that battery self-discharge is particularly harmful to batteries because the sulfate crystals that are formed are larger, and tend to surround clumps of healthy material. Solar panels are a good way to reduce, or eliminate altogether, the ill effects of self-discharge. Damage control is especially needed whenever the batteries will be inactive for more than 30 days.

Properly cared for, a battery would receive exactly the amount of self-discharge current from a solar panel. Since the sun doesn't shine full time, a solar panel can't be the perfect answer. However, by choosing a solar panel appropriately, the damage of self-discharge can be alleviated.

For each 100 Ah of battery capacity, assume a self-discharge of 50 to 100 milliamperes (mA). Convert this to a 24 hour Amp-hour number. Now choose a solar panel that will replenish this rate during average conditions for your locality.

For instance, assume that you have 400 Ah of battery capacity. This requires 200 to 400 mA of sustaining trickle current. Over 24 hours, this amounts to 4.8 to 9.6 Ah. Assume also that you get about 5 hours of sun daily. To get 4.8 to 9.6 Ah in five hours requires a solar panel output of .96 to 1.92 Amps. Obviously, with a 2:1 spread of numbers, coupled with the uncertainty of the sun, no perfect match can be made. For new batteries, choose a panel on the low side. Older batteries will require a panel on the high side. No matter how you choose, the solar panel will make a big difference in the life of your batteries.

6.8   Mounting Solar Panels

To be effective, solar panels must be aimed toward the sun. A land based solar array can make use of a solar tracker mechanism, but a boat or *RV* must usually seek other ways to achieve solar focus. One way *not* to obtain maximum output is to permanently fix solar panel location. It is our opinion that solar panels should be mounted to clamp-on brackets which may be attached to shrouds, lifelines or other convenient places. A dodger, with a main boom in the way usually has too much shade to be maximally efficient. At the price of solar panels, who can afford to waste them?

By providing a moveable mounting arrangement, you can not only achieve solar focus, as your relative position changes during the day, but you also can put your precious panels below in safety when the weather kicks up. Stowing panels is also advised in many U.S. marinas and perhaps in some foreign anchorages.

The clamp-on brackets used by portable *BBQ* units are easily adapted to mounting solar panels on lifeline stanchions. Clamp-on antenna mounts can also be used.

6.9   The Issue of Quantity

How many solar panels do you need? As many as you can afford and find space for. Everyday the sun delivers power to the earth without regard to race, religion, sex, or national affiliation. Under these conditions, we wish that everyone could obtain all their energy needs. The more energy obtained from the sun, the less you will have to run an engine, or tap the local utility.

6.10   Summary

By all means, include one or more solar panels in your energy arsenal. Don't bother permanently mounting them on *RV*s or boats ... arrange a quick clamp-on method to focus them on the sun. Connect the solar panel to charge the in-use battery, letting the other bank rest. When you must leave your batteries untended for longer than 30 days, use a solar panel to prevent deleterious self-discharge. If you install a large panel or array of panels, install a regulator to prevent overcharge. Always incorporate an isolation diode to prevent battery discharge whenever the sun is not shining.

Chapter 7

Wind and Tow Generators

7.1   General Information

Oh, the promises of free energy from the wind. Wind machines might best be operated in front of the mouths of their advocates. Yet there is energy in the wind, and it can be harnessed to charge batteries. The force of wind is proportional to the square of its velocity, and the energy which can be extracted from it is proportional to the volume and mass of air moving past the propeller. This can be stated as:

$$P = (m)(t)V^2/2$$

In the equation, mass (m) is multiplied by time (t) and the square of the wind velocity (V), to yield power. Since mass and time are also related to the wind velocity, the power derived is related to the cube of wind velocity. The mass of air moving past the propeller is related to the area of the circle swept by blade rotation. Mass of air is related to its density. Power can therefore be stated as:

$$P = (k)(d)(A)V^3/2$$

In this equation, k is an efficiency factor for the blades, d is the density of the air, A is the swept area of the blades, and V is the wind velocity.

What all this means is ...you need a strong wind and a large prop. It is our experience, that just about the time the wind is strong enough to really start producing energy, it is too strong to keep the wind machine flying. This fact can be seen from the equation above. Suppose that you are generating 10 Watts of power from 8 knots of wind. In other words:

$$10 = (k)(d)(A)(512)/2$$

From this, we can determine that the product of k,d and A equals 0.039. Now, let the wind increase to 16 knots and calculate P. The answer is 80 Watts, or 8 times the power for a doubling of wind velocity. In practice, the higher wind velocity may increase the efficiency of the generator, so the power increase may be even greater. The wind machine, then is a device that operates over a fairly narrow range of wind velocity.

But, in some harbors of the world, there is a daily breeze as predictable as the sun that makes wind. Here, a small boat swinging at anchor, and away from man made obstructions, can derive much energy from the wind. With a little practice, a propeller can be made with the correct span and variable angle of attack over the span to match the prevailing wind velocity, and the generator parameters.

The same generator used to harness wind is often used while under way by towing a water propeller. With the right prop, and forward speeds above 4 knots, much electricity can be produced. Despite the fact that the generator is now driven by the force of moving water, electricity is still produced by rotating a magnet inside coils of wire. We shall not differentiate between wind and tow generators subsequently.

While we may not seem enthusiastic about wind/tow generators, we have an old and tired unit, and have found a *new* surplus *DC* motor at an aerospace company which is a premier wind/tow machine. We definitely consider wind/tow generators to be a part of our energy arsenal.

7.2   Types of Wind Generators

There are two basic types of wind generators, converted alternators, and permanent magnet *DC* motors. An alternator is an alternator, and the chapter covering alternators is generally applicable to alternators driven by the wind. An alternator suitable for wind drive must be able to generate at a fairly low *RPM*, so don't expect much performance from a run of the mill automotive alternator. Under tow, the force available to drive the alternator can be huge, ultimately limited by sail area and the size of prop you can tow. Under these circumstances, power is available to run a transmission (belt, chain or gears) which will spin the alternator rotor at a higher speed.

The most efficient form of wind generator makes use of a permanent magnet *DC* motor. When the motor is driven, rather than driving, electricity is generated, rather than consumed. Motors suited for duty as wind generators are those which have been designed for high performance position servo systems. Examples of these include big and fast computer tape drives, as well as automatic machine tools. Unfortunately, such motors cost dearly. You had better plan on finding a used one, or buying an alternator type wind/tow generator.

Wind generators are fun to play with, and while most of the first advocates were experimenters without the technical wherewithal to extract peak performance, the market may be reaching a more stable and knowledgeable era. Whether you chose to play or buy, keep in mind that most wind generators do indeed need regulation, and/or constant vigilance. Those wind generators which have been designed to be permanenetly mounted produce very little energy, and thus do not need regulation except in areas where there is a strong and steady prevailing wind.

7.3 Wind/Tow Generator Regulation

A wind generator by nature, takes constant attention least the wind velocity exceeds its capability. In July of 1983 we witnessed the aftermath of a cyclone like storm called a *chubasco* which roared into Puerto Escondido, Mexico. It struck with hurricane force at dawn. About thirty wind generators, left dangling lifeless over night roared into action and quickly self-destructed before sleepy owners could get them down.

You may chose to regulate by the disconnect method ... when the batteries are topped off, disconnect the generator. This isn't quite as effective if you are making a passage and towing a prop. Under this circumstance, you probably are using electricity constantly, and would like to have the generator replenish that used, while not over charging your batteries.

An alternator type machine is regulated by controlling the field current, just as the engine driven alternator.

Regulating the permanent magnet DC motor is not quite as simple. It can be done with a shunt regulator, such as that used for a solar panel, but, in a real blow, a good DC motor may produce as much as 15+ Amps. This is more than a shunt regulator designed for solar panels can accommodate. Shunt regulation poses another problem to the DC motor. At a sufficiently high current, the motor windings can actually demagnetize the permanent magnet, ruining the motor. A shunt regulator works by shunting the current to ground, rather than to the batteries. It thus acts more like a short circuit on the generator, allowing much current to flow.

A series on/off solid-state switch can be used to achieve regulation. As in the case of the solar panel, the switch must have a low forward voltage drop and be able to pass the expected maximum current from the generator. Yet another problem rears its head for series regulation. When current is interrupted, the voltage from the generator rises fast and high. The switch must be protected or otherwise be able to sustain the high voltage. As you may have guessed by the lack of wind generator regulators on the market, this is a tall order.

The biggest problem with regulation of a wind generator is the fact that the unloaded generator tends to race in the wind. This can destroy a propeller which is the least bit unbalanced. A towed generator doesn't race quite as much when unloaded, but will still increase RPM considerably.

7.4   Hooking up the Wind/Tow Generator

Like the solar panel, the wind/tow generator must be connected with a diode in series to prevent battery discharge when there is insufficient

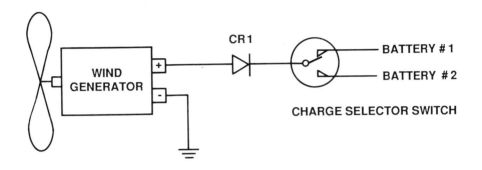

Figure 7.1: Wind Generator Hook-Up

motion to provide charging. The diode must be rated at least 20 Amps and should be mounted on an appropriate heatsink[1]. Your first choice should be a Schottky diode. Figure 7.1 shows the correct way to connect a wind/tow generator.

7.5   Summary

The wind generator can be an effective source of energy in many anchorages. A predictable prevailing wind is necessary so that a propeller of the right size and pitch can be made. Under tow, considerable power can be obtained. The wind/tow generator will require dedicated ownership, but the energy rewards can be substantial.

---

[1]Later in the book, information is presented which will allow you to determine how much heatsink surface area is required.

Chapter 8

The AC System

8.1 General Information

We owe the invention of alternating current (*AC*) to an inventor named Nikola Tesla. The unit of magnetic flux density is named Tesla in his honor. Tesla went to work for the Continental Edison Company in 1882. Two years later he immigrated from Paris to the U. S. and worked directly with Thomas Edison for a while. Leaving Edison over a dispute about payment for an invention, Tesla took up invention on his own.

The Continental Edison Company provided *DC* power to its customers and Edison himself was committed to the use of *DC* power. Meanwhile Tesla invented transformers which could step up *AC* voltage to high values so that power could be efficiently transmitted over long lines. Tesla also invented the *AC* motor, which was and still is less expensive to manufacture than the *DC* motor. Tesla teamed up with his friend and fellow inventor, George Westinghouse and *AC* in the U.S. was on its way.

8.2 What is *AC* ?

Most people know that *AC* stands for alternating current and it can shock you. That may be sufficient knowledge for a mindless consumer, but the user of an alternate energy system should know more.

We speak of current flowing in the same sense that we speak of water flowing. This is not a good analogy. It was first thought that electrons actually traversed through wires at an incredible speed ... the speed of light. Not so. Electrons in a wire move at a snails pace compared to light.

The phenomena of current flow is better illustrated by analogy to a train. When the locomotive turns on power and begins to move, all the cars begin to move also, minus a little slop in their couplings. The caboose does not *instantly* occupy the former position of the locomotive, but the *effect* of the locomotive is felt more or less instantly at the caboose. Electrons in a wire can be thought of as cars in a train. When a voltage (locomotive) is applied to a closed circuit of wire, the caboose electron feels the effect at the speed of light, though its actual position in the wire changes slowly.

DC current flows in one direction only. *AC* current flows in one direction for a while and then flows in the opposite direction. In the U.S., the direction of current flow switches 60 times a second. The *AC* line is said to have a frequency of 60 Hertz[1].

To put *AC* into our train analogy, consider a train that is used to haul ore from a mine tunnel. The train goes into the tunnel and out of it. Overall, the train doesn't really go anywhere, but in the process of going nowhere, it does much work.

AC is produced by synchronous generators. This is the same kind of machine as the *DC* alternator, but minus the rectifying diodes. Refer to Figure 8.1, and Figure 8.2. In those figures are shown a sine wave, and two views of an electro-magnetic *AC* generator. Included in the views are schematic representations of the magnetic field and stationary winding. The schematics are shown just below a quasi-mechanical view of the machinery.

Figure 8.2 illustrates a rotating magnetic field which moves *flux* through a coil of wire. In view 0, no flux *cuts* the wire and so the induced voltage in the coil is zero. In view 1, maximum flux cuts the coil and the induced voltage is at its maximum. See also Figure 8.1.

Because the generator contains a rotating field which is magnet-

---

[1] Heinrich Hertz verified the equations of Maxwell regarding electromagnetic theory, and the unit of frequency is named in his honor. Hertz is abbreviated *Hz*.

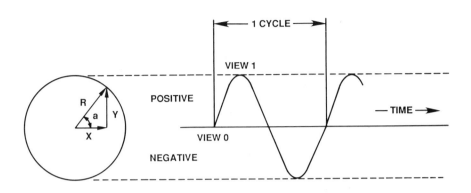

Figure 8.1: Sine Wave Generation

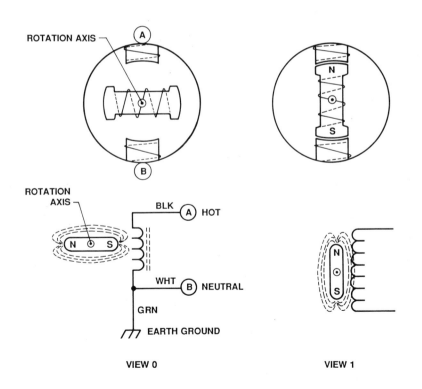

Figure 8.2: *AC* Generator Views

ically coupled to stationary windings, the output voltage is a sine wave. That is, the instantaneous voltage at any time is proportional to the trigonometric sine of the angle between the rotating field and the stationary windings. The rotor is excited by DC current and thus has a magnetic north pole and a magnetic south pole. The north pole produces current in one direction as it passes the stationary winding while the south pole produces current in the opposite direction.

In Figure 8.1 is shown a circle with an enclosed triangle. The value Y/X is the trigometric sine of the angle a. See also Figure 8.2. X represents the passage of time, where Y is the magnitude of the voltage produced when a magnetic field rotates inside a coil of wire. One cycle of the sine wave is one revolution around the circle. Magnitude Y is equal to the trigonometric sine of angle a. View 0 of Figure 8.2 shows the magnet position relative to the windings when the generator output (Y) is zero. Note that the magnetic flux from the electro-magnet does not *cut* the stationary windings.

When the magnet is rotated as shown in View 1 of Figure 8.2, then magnetic flux does cut the winding and maximum output is produced. As the magnet rotates another 90 degrees, then no flux will cut the winding and the output will be zero again. Another 90 degree rotation will bring the south pole to the top. Again, maximum output will be produced, but this time current will flow in the opposite direction because the magnetic poles are swapped from that shown in View 1 of Figure 8.2.

As the generator field rotates, the output voltage swings through zero, maximum positive, zero, maximum negative and so on. One cycle is shown in Figure 8.1. As noted above, the unit of frequency is named Hertz. Sixty *Hz* is equivalent to 60 cycles per second, the common frequency used in the U.S. *AC* is the *train that goes nowhere*, but does much work in the process.

8.3   Safety First

Use extreme care when working around *AC* voltage. Never do any wiring on *hot AC* circuits. When troubleshooting *AC* circuits it is a good practice to wear insulated shoes so that circuit paths cannot be made through your feet. In addition, never use both hands to grasp wires or equipment because you may grab both sides of the *AC* line. It only takes a few micro-Amps[2] to kill in such case. *AC* voltage also has a *gripping* effect which can prevent you from releasing the wires. Have a friend nearby to shut off *AC* power should the need arise.

AC doesn't really grip ... the constant reversal of current weakens muscular control by interfering with normal nerve activity. Current from one hand to the other flows through the heart.

(Contact with *DC* results in a repulsive reaction which tends to disconnect you from the circuit. We were *blown* off a lab stool by 2500 Volts *DC* years ago, and received a large bump on the head from hitting the wall behind. Did we learn respect? You betcha!)

AC is brought to us from the power company over three wires. The black wire is the *hot* lead, relative to the other two. The white wire is called the neutral wire, while the green wire is called the safety wire. At the power plant, the green and white wires are connected together. You should never connect the white and green wires from the power company ... leave that to them. However, when you install your own *AC* generator at a remote home site, be sure to give it a good ground connection. Bury several feet of large diameter construction rebar into the ground and connect it to the chassis of the generator with heavy wire. Choose a sight where the ground is naturally moist. The life you save may be your own.

8.4   How Much *AC* Do You Need ?

As we have indicated earlier, *AC* cannot be used abundantly in an economical energy system. It isn't that *AC* itself is inefficient, but rather that many appliances which run on *AC* are. The most notable offender is the household refrigerator and freezer. To live economically from an alternate energy system, reliance on *AC* must be avoided. Before buying an *AC* generator, consider the use of an inverter for infrequent *AC* needs. Consider also our suggestions regarding the *DC* alternator and small engine, presented in a preceding chapter.

---

[2]One micro-Amp is 0.000001 Amp. Several micro-Amps through the heart muscle can be lethal.

Despite our reluctance to use *AC*, there are appliances which we operate that require it. These include the computer being used at this moment, and the many power tools used for construction and repair on our sailboat. All of these are operated from an inverter. (Inverters are discussed later.)

In other situations, washing machines, dryers, microwave ovens, large televisions, stereo systems, and ventilation may require larger amounts of *AC*. Inverters to run loads greater than about 2000 Watts are available, but they must have 24 Volts or more to operate. The source for large *AC* loads must be an engine driven generator. Sizing the generator that is necessary involves the tabulation of wattage requirements for each apparatus and then determining how many of them need to run at one time. Given in Table 8.1 is the power required by various appliances and tools.

By choosing different times that each appliance is operated, you can avoid the initial and the operating expenses of a large generator. You may elect to use a 3–5 kiloWatt *kW* generator at various times during the day. Remember that a generator has an engine which should be warmed up before placed under heavy loads. Remember also that the heating and cooling which goes along with starting and stopping an engine can be more detrimental to an engine than constant operation. For best results, schedule the daily chores so that all your *AC* needs can be met with one running of the generator.

With the proper battery system and a high power inverter, you can likely make do with a few hours a day from a 650 Watt generator. The generator is used only for charging batteries. The inverter must be able to supply the necessary *AC* power. Even if you can get by on a small generator and a large inverter, you should still try to schedule the use of large *AC* loads at a time when the generator is running and charging the batteries. At least some of the input power to the inverter will come from the generator instead of from the batteries.

When you size *AC* loads, you will discover that *AC* appliances are either specified in Watts or *VA* (Volt-Amperes). For *DC* circuits, Watts is equal to the product of Volts and Amperes. In *AC* circuits, Watts is not always equal to *VA* due to a phenomena called power factor (*PF*). *AC* motors, it turns out, draw maximum current some

Appliance/Tool	Watts
Air Compressor (1 *HP*)	1500
Air Conditioner	900–∞
Blender	300
Coffee Maker	1000–1200
Computer	65–200
Drill, Hand (1/4)	300
Drill, Hand (1/2)	600
Hair Clippers	75
Hair Dryer	1000–1500
Iron	1000
Micro–Wave	900–1500
Printer	150–400
Razor	30
Refrigerator	500
Saw, Sabre	500
Saw, Circular 7–1/4 inch	1000
Sewing Machine	250
Stereo	70
Slide Projector	600
TV, Small B & W	80
TV, Color	300
Typewriter	250
Vacuum Cleaner	600–1200

Table 8.1: Common Appliance Power Requirements

time after the peak AC voltage[3]. Since your principle AC loads will be motors, a bigger generator will be required than that indicated by the Wattage rating. For large induction motors, doubling the expected wattage rating will be required. For instance, a 1 HP motor develops 1 HP which is equivalent to 746 Watts. Since the motor is only 70–75% efficient, input power to the motor will be about 1000 Watts. Because of the power factor, however, a 2000 Watt generator will be required to start the motor and run it under full load.

Power factor is the ratio of two power calculations, one involving Volts and Amps as if their peak values occurred simultaneously, and the second involving Volts and Amps taking into account the phase difference between the peak values. A motor with a power factor, (PF) of 50% will take twice as much power to operate as a water heater coil of equivalent energy rating. A heating coil is mostly resistive in nature, so its power consumption is simply the product of Volts times Amps.

Besides needing more power to run, induction motors also need alot of current to get started. The generator must be able to supply the starting current. Some generator manufacturers specify the size of motor that can be started by their products. We compliment them on forthright specification, and highly recommend that you only purchase a generator that carries such a specification.

Note that most small power tools use so called *universal* motors. Universal motors run on either AC or DC. They do not demand large starting currents and have power factors close to 100%. Doubling their rating is not required when calculating generator or inverter size. A 1000 Watt inverter is sufficient for most small power tools. We use one for our vacuum cleaner, power saw, sander, large drill motor, soldering iron, wheat grinder, and small table saw. Our computer and printer, and smaller tools operate from a 300 Watt inverter.

---

[3]An AC motor is principally an *inductive* load, and therefore the current flow lags the applied voltage. The lag in a pure inductor is 90°, but motors are not pure inductors.

8.5 Auxiliary Generator Adjustment

The auxiliary generator should be adjusted to operate at the correct frequency and produce the correct voltage at that frequency. Frequency is controlled by the *RPM* at which the generator runs. The higher the *RPM*, the higher the frequency. To adjust frequency requires a tachometer or an oscilliscope which displays the *AC* waveform. Voltage can be adjusted using a sensitive digital voltmeter. While frequency is not usually as critical as voltage, both should be set as a periodic maintenance item.

You should note that ferro-resonant battery chargers are frequency sensitive. A frequency higher than 60 *Hz* will defeat the regulation characteristics of such chargers. If you drive the charger with too high a frequency it will overheat and fail. With a little judicious adjustment of generator *RPM*, however, you can squeeze a little more charging performance from a ferro-resonant charger. In particular, you can shift the output cutoff point higher, allowing a faster charge. To do so, shift the frequency higher, to the range of 65–70 *Hz*. This trick is not for the inexperienced ... no charging is worse than slow charging, and there is a fine line between *OK* and *smoke*.

8.6 *AC* and Boats

There are two major issues of *AC* which are unique to energy systems on boats. Safety and electrolysis. These subjects are covered in a separate chapter. Not covered in that section is the proper wiring procedures for *AC* in general, which are presented below.

8.7 *AC* Wiring

Figure 8.3 shows the wiring of an *AC* system. Shown are two sources of *AC*, the power utility and an *AC* source of your own. It might be an auxiliary generator or an inverter operating from your *DC* system. A transfer switch is wired to select which *AC* source you use. As shown, there are only two positions on the transfer switch. You might consider a three position switch, which would allow selection of utility power, auxiliary generator or inverter. Generally this is not required. An alternative to the 3 position switch is to make a cable that connects from the auxiliary to the power receptacle that is normally used to plug in shore power. When you aren't tied up to the utility, plug in the auxiliary.

Beyond the transfer switch is shown a polarity indicator. These devices are a *neon* bulb and series resistor. As shown, the lamp should not light because the white wire and green wire are supposed to be at the same potential ... earth ground. If the light ever does light then the black and white wires are swapped somewhere. Disconnect from shore power immediately, and check out wiring starting with the shore power receptacle. You will need an *AC* voltmeter to measure the power connections in the receptacle. Measure the Volts from the individual wires to the earth ground safety wire. You need to mate your white wire with the receptacle wire which measures zero (or real close to it).

If you are wiring a new *AC* system, be sure to put the polarity indicator between the white and green wires. Don't be clever like one person we met who had wired the indicator between the black and green wire. His theory was simple enough ... the light should glow, indicating the presence of *AC*, as well as the correct polarity. What he failed to consider, was the small, and constant current which was returning to ground via the safety wire. How many of his neighbors zincs had his indicator cost?

Following the *AC* transfer switch is the Main Circuit Breaker. We do not recommend using fuses because they can be dangerous to change, may corrode in place, and your spare probably can't be found when it is needed. Circuit breakers are a better choice. Make sure the main breaker is rated for all of your intended loads. For boats, use either a 30 Amp or 50 Amp breaker. (You should always carry a spare breaker or two. When buying breakers, always get the combination *AC/DC* types. They cost a little more, but can be used in either the *AC* or *DC* circuits.)

In Figure 8.3 next to the polarity indicator, is a ground isolator. This circuit is discussed in the chapter covering *AC* and Electrolysis. Its purpose is to prevent electrolysis while providing a connection between internal *DC* ground and earth ground. With the isolator

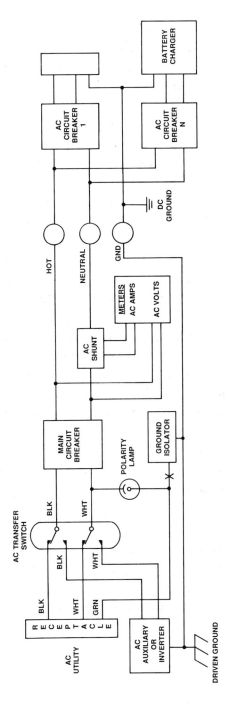

Figure 8.3: *AC* Electrical System

installed, it is good practice to verify that no current flows between the two grounds. Break the circuit at the point marked with an X and measure current with an ammeter. Try both AC and DC scales ...expect no current on either scale. Don't forget to re-connect the circuit after measurement. Rather than break the circuit each time you verify no leakage, you can permanently install a shunt, such as a 50 milli-volt, 50 Amp shunt. With the shunt in place, current can be measured with a good digital meter. You will need sensitivity of about 10 micro-Volts per digit.

Your AC system should most certainly be instrumented with both AC Volts and AC Amp meters. Better yet, use isolated input sensors to digital meters for measurement of Volts and Amps. AC motors can be quickly destroyed when run from AC service with too high or too low a voltage. The AC motors in hermetically sealed refrigerators are very prone to failure when the line voltage is not correct. A little knowledge can save an expensive overhaul.

Knowing AC current consumption at any time can save those annoying occasions when dockside circuit breakers trip, or your auxiliary generator groans under an overload. Before you turn on that AC appliance, check the current draw to see if there is enough power left to run the appliance. Perform the same check after the appliance is running. AC generators are too often abused by uninformed users, and fail prematurely as a result.

As shown in Figure 8.3, AC Volts are measured across the lines, from the black wire to the white wire. To measure AC current, a shunt is wired in series with the white neutral wire. You can wire it in series with the black wire ... placed there it would register correctly even if the current was returning via the safety wire. Under normal circumstances however, putting the shunt in the neutral side will be safer because there will be one less connection to the hot wire. Placed in the white wire, the measurement can also indicate when current is returning via the green safety wire. Any time you have a known AC load, and there is no AC current indication, the current is returning via the safety wire.

At some convenient location, bring your AC system to three protected terminals. From here, and only from here, string the individual wires to circuit breakers which feed appliances and outlets. Be sure to use 3 wires per circuit. As shown in Figure 8.3, the safety wire connects to your *DC* ground. With the isolator intalled properly, you will be protected from leakage currents and suffer no electrolysis.

Keep all your circuit breakers in one central location and well protected from water or other moisture. When a breaker trips, you will know where to find the breaker, and by keeping the breakers centralized, you will have fewer unprotected *AC* lines strung around.

The wire you use for the *AC* wiring must not only be sized properly, but it should also tolerate the damp environment which is typical of a boat. Do not use conventional house wiring. Instead, use 3-wire cable rated for use outdoors, or rated by the Coast Guard for marine service. Do not string *AC* wires through the bilge even though you use good cable. If possible, route *AC* wiring highup, just under deck level.

Second thoughts should be given to *AC* outlets in wet area such as heads or bilges. If there is any question about the safety of an *AC* outlet, use a ground-fault isolator (*GFC*). A *GFC* is a differential current sensing device that senses the current in the black and white wires. If, at any time, the two currents are not equal, then the *GFC* opens the circuit. They operate fast enough to save lives.

Go light on *AC* service, reserving outlets for bare essentials. How to choose wire sizes is covered in the Appendix.

8.8 Inverters

Inverters are devices for converting *DC* power into *AC* power. (Note that the word converter is applied to equipment which converts *AC* to *DC*, as does the conventional battery charger.)

A simplified schematic of an inverter is shown in Figure 8.4. At the heart of the inverter is a split primary transformer. The transformer has more secondary turns than primary turns, thus it steps up the voltage applied to the primary. The center tapped primary is driven by two Solid-State-Switches, (SSS) such as transistors or MOSFETS. The switches are controlled by a 60 *Hz* oscillator. The oscillator alternately turns on the two switches. When SSS1 conducts, one polarity of *AC* is generated. During the next half cycle, SSS2 is turned on and

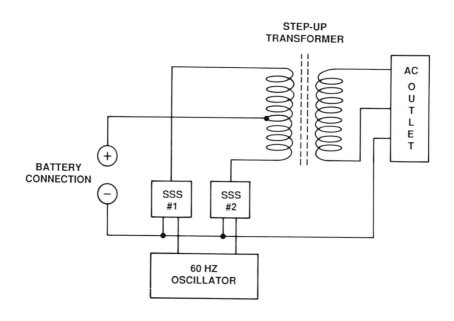

Figure 8.4: Square–Wave Inverter

the other *AC* polarity is produced. As long as the oscillator runs, *AC* is produced at the oscillator frequency.

Oscillator frequency is important for some *AC* appliances which operate with synchronous motors. For instance, stereo turntables require a precise 60 *Hz*. Television and some computer terminals like to have a precise 60 *Hz* ... without it you may get screen jitter. Most other appliances are not as sensitive to frequency with the exception of the ferro-resonant battery charger. It is unlikely that you will be running the battery charger from an inverter. If you have frequency sensitive appliances, then be sure to choose an inverter with a crystal controlled oscillator. A crystal provides a more precise frequency which is also more stable over time and temperature variations.

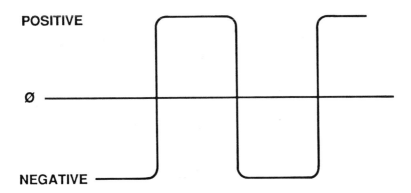

Figure 8.5: Square Wave

8.8.1 Inverter Output

Inverters come in two basic varieties. Those that produce a square wave and those that produce a sine wave. As you know, the *AC* power supplied by generators is a sine wave.

A square wave is shown in Figure 8.5. Basically a square wave has values of either maximum positive or maximum negative. The transition from one maximum to the other is almost instantaneous. The corners of the square wave are shown somewhat rounded in Figure 8.5. In practice, the square wave output of an inverter will have much more rounded corners than shown. That's because the transformer in the inverter is not capable of switching instantaneously from one maximum to the other. As it turns out, this is an advantage rather than deficit. The leading and trailing edges of a true square wave are rich in high frequency harmonics. A typical *AC* load would have to dissipate the power in those harmonics as heat since no useful work can be done by them. Wasted energy cannot be tolerated, and the excessive heat generated may be hazardous to your equipment.

Square wave harmonics can also cause television and stereo noise. The best cure for this type of problem is to get *DC* operated equipment and dispense with the inverter all together. As stated earlier, a practical alternate energy system uses *AC* only when necessary. If

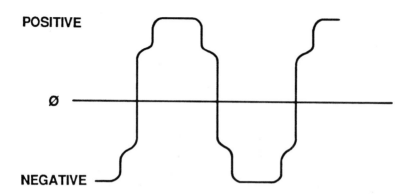

Figure 8.6: Stepped Square Wave

equipment can be powered directly from the batteries, it hardly makes sense to incur inverter inefficiencies.

Figure 8.6 shows a stepped square wave. As shown, two steps make up the transition from one maximum to the other. Two steps are not commonly used, but there are inverters which use 3 steps to make the transition. By using stepped drive to the transformer, the output can be made to more closely approximate a sine wave. In practice, the steps shown in Figure 8.6 would not be as pronounced. As before, the limited frequency response of the power transformer helps to smooth the steps into a more linear slope.

You probably won't see many square wave inverters advertised. As noted, the transformer will not generate a true square wave, but instead rounds the corners. Manufacturers use this fact to call their product a *modified square wave* or a *quasi sine wave*.

In Figure 8.7 we show an exaggerated example of a synthesized sine wave. Such an output is just a further extension of the stepped square wave shown in Figure 8.6. Instead of two or three steps, the output of Figure 8.7 is comprised of many small steps. The number of steps through the cycle may be many thousand. The average value of the steps is that of a pure sine wave. Again, the waveform of Figure 8.7 does not represent reality. Because the output is filtered, the

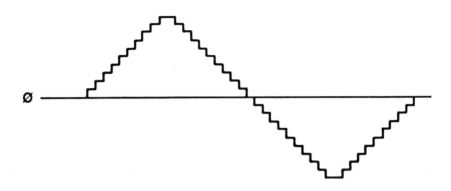

Figure 8.7: Synthesized Sine Wave

individual steps merge and very closely approximate a true sine wave. How close the approximation is depends on the number of small steps and the effectiveness of the filter. Whereas the square wave inverter is relatively simple to design and build, the synthesized sine wave inverter is complicated and thus expensive.

An inexpensive square wave inverter is not dynamically regulated, that is the output is not sensed and feedback corrections made to the input. For that reason, output voltage will be dependent on the battery voltage. As the battery starts to decline, so will the inverter output.

The sine wave inverter must contain feedback circuitry to measure the output voltage and adjust the size of the input steps to keep the output at the desired value. Within limits, the sine wave inverter can compensate for battery voltage as it discharges. Such feedback compensation is not without its price, however. Under certain combinations of input voltage and output load, the feedback loop becomes unstable. When that happens the inverter may just cook whatever load is attached to it, as well as itself.

8.8.2   Inverter Rating

Inverters are rated in Watts, but despite that, look out for how the watts are specified. The most meaningful specification will be the continuous *RMS* Watts which the inverter will stand up to. In addition to that, some manufacturers specify a peak wattage rating for some limited duration. This number is significant if you want to use a small inverter to drive a universal motor for a short duration. Such motors don't require large starting currents, so a small inverter with a large short-term overload capacity will suffice to drill a hole or make a cut, even if the motor rating is for more Watts than the continuous inverter rating.

The specification that is missing from most inverters is the size of synchronous motor that can be started. Will the 1000 Watt inverter start that 1 *HP* motor, or do you need to buy the 2000 Watt model? Only the manufacturer can tell you for sure.

Another specification that is usually missing is the permissible inrush current allowed. Resistive loads such as lights, toasters and water heaters require enormous current until the element gets hot. A light bulb, for instance has an inrush current about 7 times its steady-state value.

As noted above, certain conditions of input voltage and the load on the output can cause instability in sine wave inverters. In particular, induction motors draw peak current long after peak voltage is applied. This is because of the large inductance of the motor windings. The effect can be balanced by adding capacitors across the motor, but the inverter may not operate with either the large inductance of the motor, or the capacitor compensated combination. We would not buy a sine wave inverter unless it carried a specification for such conditions. Look for the specification, *stable under all line and load conditions, including capacitive or inductive power factors to 30%*. Such a specification will cover inrush loads such as capacitor start networks on motors as well as the inductance of an uncompensated induction motor.

Because a synthesized sine wave is made up of many small steps, there are considerable high frequency components in the generated waveform. The output will be filtered to prevent conduction of these high frequencies to your *AC* loads, but there may still be enough radiated energy to interfere with radio and television equipment. Don't be surprised if your favorite program can not be received when the sine wave inverter is running. Under *FCC* laws[4], inverters are required to pass stringent electromagnetic interference tests, but it is unlikely that qualifying tests have been performed by the manufacturers.

Don't be mislead by meaningless statement of fact. It is a fact that the instantaneous power at the peak of a sign wave is double the *RMS* power of a sine wave. So what? Instantaneous peak power is meaningless. *RMS* power is real power.

8.8.3 Inverter Efficiency

Efficiency is stated in percent of input power that is delivered to the load. If an inverter is consuming 100 Watts of battery power and delivers 90 Watts to the load, then it is 90% efficient. Inverter efficiency varies from good to dreadful. The simple square wave inverter shown in Figure 8.4 will have poor efficiency when no load is attached because it takes some amount of energy just to magnetize the transformer. Often the simple square wave inverters are made with *bipolar* transistors which require considerable driving energy ... the result is inefficiency. MOSFET transistors are used in more modern (and expensive) inverters. MOSFET devices have very low series resistance in the on-state, and extremely low control energy, thus less power is wasted.

To lower the power required when there is no load on the inverter, some manufacturers sense the risetime of the output and shut off the solid state switch if no load is sensed. This results in short pulses at 60 *Hz* rather than a square wave. When a load is sensed, then the pulses are stretched to deliver more power to the load. Under full load, the output is a full duty cycle waveform.

Manufacturers specify efficiency at the best load conditions. Usually that will be with a load which is 30–60% of the maximum load. Before buying an inverter, get a full efficiency curve which shows efficiency under all load conditions.

---

[4]*FCC* Rules, Part 15, Subpart A attempts to regulate any device which can radiate energy from 10 kHz to 3 GHz.

You should also check out the no–load draw of the inverter. How much energy does the inverter use when it is turned on, but with no load attached to it? When you are intermittently using power tools it is handy to let the inverter run rather than turning it off each time you finish a hole or a cut. Using an inverter to run a sewing machine can also be inefficient unless the inverter has little current draw between stitches.

8.8.4   Which Inverter to Buy

The square wave inverter has been around for some time and as a result their cost is generally much less than sine wave counterparts. Motors operate best on a sine wave because they use rotating magnetic fields just as the AC sine wave generator. On a square wave, motors tend to *knock* and under heavy loads will heat up more than when run by a sine wave. This isn't usually a problem because inverters are not used for long-term power to motors under heavy loads. Typically, the inverter is turned on briefly to drill a hole or cut a piece of lumber.

Equipment which contains internal power supplies generally work as well with a square wave input as with a sine wave. This includes computers and printers. Water heaters and toasters also work as well from a square wave, even using the harmonics as heat energy.

For limited usage, the square wave inverter is economical and quite functional. Our computer, printer, and small tools work well on an efficient 300 Watt unit. For larger tools we have an old, tired, noisy, inefficient but still functional 1000 Watt inverter that tolerates huge short-term overloads and appears to be indestructible.

For those with microwave ovens and a wider range of AC usage, then the synthesized sine wave inverter should be considered.

8.9   Summary

This chapter has dealt with AC within an alternate energy system. The generation of AC and its sine wave characteristics have been presented. Selection and use of a small generator has been discussed. The types and characteristics of DC to AC inverters have also been given.

Chapter 9

Principles of Refrigeration

9.1 General Information

Most of us regard refrigeration as a basic necessity. On a personal level, refrigeration preserves our food and cools our beer. On a larger scale, refrigeration is used throughout industry in many processes. Medicine as we know it could not exist without the aid of refrigeration. If and when you venture south, read the storage conditions on the medicine you take. Much of it should be stored under refrigeration.

Given the great benefits of refrigeration, it is only natural for us to demand it from our alternate energy systems. However, it won't come cheaply. Refrigeration will be the major energy consumer in your system. (We intentionally neglect heating requirements.)

Because refrigeration will be the major consumer, choosing the right refrigeration technology, and using it optimally will pay the greatest dividends. Choosing the technology and using it to your best advantage will not come about by buying all that the refrigeration salesperson tells you. You will need to know more about heat in general, how it is extracted, and how to keep something cold after the heat has been removed. To paraphrase an old truism, we might say *knowledge is cold.*

9.2   History of Refrigeration

The early Egyptians enjoyed the benefits of refrigeration. At night they put wine into porous earthenware and placed the containers on their rooftops where the night air and the natural process of evaporation cooled the contents. Chinese, from the era of first documented history, harvested crops of ice for use later in the year. Ice was a major international commodity until the invention of ice making equipment.

The manufacturing of ice was made practical in 1834 by Jacob Perkins, a U.S. inventor, who developed the compressor technique. The principles behind absorption refrigeration were discovered by Michael Farady in 1824, and applied in 1855 by another German inventor.

A warm winter in 1890 caused a worldwide shortage of ice and sparked the mechanical ice making revolution. By the early 1900's, domestic refrigeration became available. The first automatic domestic refrigerator was introduced in 1918 by Kelvinator[1]. They sold 67 units the first year. General Electric[2] sold the first *hermetically* sealed automatic refrigerators in 1928. Today, more than 10 million units are sold annually.

9.3   Breaking Bad Refrigeration Habits

Before we delve into the technical aspects of refrigeration, a look at how to use refrigeration sparingly is required. As we noted above, the cost of refrigeration in the alternate energy system is high. No matter how you attain refrigeration, your cost for it can be reduced by intelligent usage.

The first rule is simple. Don't refrigerate it unless it needs to be refrigerated. When you are connected to the power pole, and have the latest in domestic refrigeration equipment at your disposal, everything seems to migrate into the box. This is a tough habit to break when a large refrigerator suddenly becomes impractical. Most fruit and vegetables will last for several days in a cool, dry spot. If

---

[1]Kelvinator, Inc./White Consolidated Industries.
[2]General Electric Company.

you buy a stock of foodstuff, leave the first few days consumption outside the box. Before anything goes in the box, get in the habit of asking yourself, does it really need to be refrigerated.

The next rule is just as simple. Keep the box just cool enough to preserve its contents. Avoid the urge to have things ice cold. Table 9.1 indicates the best long-term storage temperature for some foods. Since you won't usually keep things for the maximum length of time, adjust the temperature upwards by several degrees. If you decide to experiment, keep good records of temperatures and the effects.

As you will note from Table 9.1 there is a different temperature required for different foods. This leads to the next rule ... keep your box organized according to the temperatures best suited for the foods. The foods which require the coldest temperature should be located in the coldest place. Typically this will be at the lowest point in the box. You can also make a small box close to the evaporator. Even cardboard can be used to gain a lower temperature nearby the evaporator, while still keeping the rest of the box cool.

Keep your box humid. To be sure, bacteria grows best in humid environments, but very slow if the temperature is less than $40°F$ $(4.5°C)$. The humidity in the box will promote a more even distribution of temperature, and prevent moisture loss from fresh vegetables and fruit. An open container of water will help keep humidity high. Melting ice serves a double purpose, cooling and humidifying. In a box with melting ice, fresh vegetables last three to four weeks.

Don't fill the box so full that the air cannot circulate. You may wish to pack some foods around the evaporator, such as smoked fish or cheese, but to keep the rest of the box cool requires some circulation. If circulation is extremely poor, a very small fan can be used. It should be a fan which barely rotates on 12 Volts, and has minscule current draw. Remember, the fan generates heat when it runs.

Keep the box relatively full, even if with plastic water bottles. The air in the bottles won't spill out when the box door is opened. If you have a large box suitable for long ocean passages, but are not using all of it normally, cut chunks of styrofoam and further insulate the box. You can hold them in place against the walls with double back masking tape.

Always put things in the box when they are the coldest. Usually

Food	Temperature Range (Faharenheit/Celsius)	Storage Time
Fresh Beef	(32–34)/(0–1)	1–6 weeks
Fresh Pork	(32–34)/(0–1)	3–7 days
Ham, Bacon	(32–40)/(0–4)	2–4 months
Fresh Fish	(32–40)/(0–4)	5–10 days
Dried Fish	(23–25)/(-5)	2–3 months
Smoked Fish	(32–40)/(0–4)	10–20 days
Ripe Tomatoes	(45–50)/(7–10)	7–10 days
Potatoes	(32–50)/(0–10)	5–12 months
Onions	(32–35)/(0–2)	6–8 months
Dry Garlic	(32–35)/(0–2)	6–8 months
Cucumbers	(45–50)/(7–10)	10–14 days
Topped Carrots	(32–35)/(0–2)	4–5 months
Berries	(31–32)/(0)	7–10 days
Oranges	(34–38)/(1–3)	8–10 weeks
Apples	(32–34)/(0–1)	2–7 months
Dried Fruit	(26–32)/(-3–0)	9–12 months
Dried Nuts	(32–50)/(0–10)	8–12 months
Eggs	(32–34)/(0–1)	8–10 months
Milk	(36–38)/(2–3)	5–7 days
Cheese	(32–34)/(0–1)	30–60 days

Table 9.1: Food Storage Temperature and Time

this will be in the morning. Every little bit helps. When you are operating from an intermittent system (explained later) add items to the box when you are pumping down the box, using the generated refrigeration rather than the hold over cold.

Don't leave things out of the box longer than necessary. If it warms up, it is only going to cost you to cool it again. Don't bring the whole jar of mayonaise to the table and let it warm up during the meal!

Don't drink cold water! Cool water will quench the thirst as well as cold water, and is actually easier on the body. Experiment by mixing the cold water from the box with room temperature water. We have discovered that about 25% cold water with the remainder at room temperature is quite pleasing. (You will probably have a hard time convincing guests that refrigeration is such a scarce commodity until they have had the experience of dead batteries and no ice for their favorite refreshment.)

Use a transfer box. A transfer box is a well insulated box that has no active refrigeration. The items which are to be consumed in the near future are placed in the transfer box with items which are destined for the refrigerator. Used in this manner, the items to be consumed absorb some of the heat from the items which are to be refrigerated. The result is less of a load on your refrigerator system. A transfer box is particularly effective for a large number of beer drinkers on a small yacht. (Editors note: This also works well for a large beer drinker on a small yacht.)

If you have a freezer, always thaw frozen food in the refrigerator.

Don't open the box more than absolutely necessary. The transfer box described above is a good way to avoid many openings. Sometime before meal preparation, remove all of the items to be consumed at one time, putting them in the transfer box. The wine to be served with the meal can be adequately chilled by adding it to the transfer box with the cool food taken from the refrigerator.

Make sure that your box is well sealed. Any small air leaks can lose a great deal of precious energy. If there is a drain in the bottom of the box, put a small elbow in the outlet, and keep the elbow full of water. This will make a water trap, preventing direct airflow. Insulate the elbow, so that the cold water in the trap stays cold.

When you have excess energy, such as when you are running an engine, put several bottles of water in the box and *pump* them down. They will store some cold. If you have a freezer, then make ice when you have plenty of energy, and transfer the ice to the refrigerator when needed.

Keep the freezer defrosted. Frost makes a great insulator, and thus cuts the efficiency of the evaporator significantly. Defrosting is required more often in systems without holdover plates, since the evaporators are usually operated below freezing and collect frost.

9.4   Definitions and Concepts

Learning about a new subject requires learning the *jargon*. Jargon is the names of entities making up the subject, and the relationships which hold true between the entities. It is always confusing at first, and only after many readings and a good deal of reflection does the jargon take on meaning. We have attempted to be as non-technical as possible and still impart the essential knowledge which is required to understand basic refrigeration principles. Nevertheless, some technical terms and the underlying principles behind them must be understood.

9.4.1   Temperature and Heat

Temperature is the intensity of heat. Water boils at a temperature of $212°F$ $(100°C)$. Heat is something else again. We can easily place our hand in an oven of $212°F$, but even the quickest immersion in water at $212°F$ would cause severe burning. The hot air in an oven cannot quickly transfer a great amount of heat to our hand, because air does not contain much heat energy. Hot air and hot water, both at $100°C$ have the same intensity (temperature), but contain vastly different amounts of heat.

Heat cannot be created or destroyed, only transferred. The transfer of heat is always from a hotter object to a colder object. A cold object is one whose heat has transferred to yet another colder object. A refrigerator is simply an apparatus to transfer heat from the inside of the box to the outside.

9.4.2 Quantity of Heat

Whereas temperature is the intensity of heat, the quantity of heat is slightly more complex. To know the quantity of heat in any given substance requires knowing *three* parameters.

• the temperature of the substance

• how much the substance weighs

• what the substance is

Heat therefore is measured in degrees per weight of specific material.

Heat is measured in *Btu*'s (British Thermal Units) or Joules. The *Btu* is the outdated unit of heat and the Joule is the International Standard unit. For the most part, we will stick with units in *Btu* since there is so much of refrigeration technology expressed in these terms already. Conversion from *Btu* to Joules can be done by the reader using the Appendix. A *Btu* of heat is the quantity of heat required to raise one pound of water one degree Fahrenheit. A Joule is the quantity required to raise one kilogram of water one degree Kelvin[3].

Notice that the definitions for *Btu* and Joule include the three parameters of temperature, weight, and a specific substance. Both use water as a reference substance, and specify the weight. Raising temperature requires adding heat, and lowering temperature requires removing heat. Both are measured in *Btu* or Joules.

Heat can be translated to mechanical energy such as horsepower and it can also be translated to electrical energy. This enables us to calculate the cost of refrigeration in terms of battery Amp-hours. For the alternate energy system, the Amp-hour is universal currency.

9.4.3 Specific Heat

What about the quantity of heat in something other than water? To determine that, we have to know how much heat the other substance

---

[3] A degree Kelvin is the same magnitude as a degree Celsius, but where as $0°C$ is the freezing point for water, $0°K$ is the temperature where molecular motion ceases. Only the offset is different between Celsius and Kelvin.

will hold, relative to the same weight of water. The ratio of other substances to water is known as *specific heat*. For instance, the specific heat of alcohol is 0.6. That means that a kilogram of alcohol will hold only 0.6 the heat that a kilogram of water will. Specific heat for materials other than water have been developed by empirical measurement and related to water as a baseline reference. The Appendix contains specific heat for some common materials.

9.4.4   Thermal Conductivity

Another term of interest is the thermal conductivity of a substance, that is to say how fast heat can be conducted into or out of a substance. Thermal conductivity is dependent on 5 parameters listed below.

- the cross sectional area (A)

- the thickness of the material (T)

- the temperature difference between the two sides of the material (dt)

- the time duration of the heat flow (Tm)

- a *specific factor* related to the material itself (k)

The quantity of heat is usually designated by the letter Q. As noted above Q is measured in Joules or *Btu*.

Tables for k factors of common materials have been developed. These factors are read as *Btu* per hour per square foot per inch thickness per degree Fahrenheit of temperature difference. Alternately, *Joules per second per square centimeter per centimeter of thickness per degree Kelvin temperature difference*. In the International Standard system, a Joule per second is one Watt, and thus thermal conductivity is usually stated in Watts.

9.4.5   Sensible Heat

Would you believe that sensible heat is heat that makes sense to the senses? If we light a fire under a pan of cold water, it makes sense that the temperature of the water will rise. In other words, we can measure the temperature and find it increasing as heat is applied.

9.4.6 Latent Heat

Latent heat is heat that goes into a substance without a corresponding increase in temperature. Water boils at $100°C$. No matter how much heat we apply to water at $100°C$, the temperature stays the same. The water will boil faster, but the temperature will remain $100°C$. The word latent means hidden ... the heat we apply to boiling water is *hidden* from the thermometer because the thermometer doesn't rise. Yet it is energy that must be accounted for. Latent heat is the energy that is required to cause a change of state of a substance. In the case of boiling water, the state of water is changed from a liquid to a vapor at the same temperature, $100°C$. (At this temperature, water vapor does not conform exactly to the so called *gas laws*, though it seems to be a gas.)

Science still stumbles on the mechanisms at work within a phase transition from one state to another, so don't feel bad if your own intuitive senses are affronted. Remember however, that latent energy is the amount of energy consumed or released by a substance undergoing a change of state. A pound of ice requires the addition of 144 *Btu* of heat energy to melt. Conversely, a pound of water at $32°F$ must have 144 *Btu* of heat removed to become ice at the same $32°F$.

The idea of latent heat as opposed to sensible heat is a key concept in the refrigeration system. How latent heat is applied to refrigeration is presented in subsequent paragraphs.

9.4.7 How Refrigeration Works

One of our difficulties in learning a new subject has always been the seemingly endless details that must be learned before any of the details can be applied. Before continuing with endless details, it is instructive to apply what is now known.

Evaporation is a process that requires a source of heat. Furthermore, evaporation is a change of state, requiring more heat than is sensible. Some liquids evaporate at very low temperature, and the heat that drives their evaporation process is therefore not *hot* by human senses. Refrigeration works around the idea of evaporating a liquid at a very low temperature, extracting the heat necessary for the evaporation process from nearby objects.

9.4.8   The Heat Trip

Specific heat of ice is about half that of water. Say we start with a pound of ice at -20°F. To make it melt at 32°F requires a temperature rise of 52°F. Thus, 26 *Btu* is required to warm the ice to its melting temperature. (One pound of material that has a specific heat one-half that of water, raised 52°F. Recall that water absorbs 1 *Btu* per pound per °F.) As we noted above, it takes 144 *Btu* to melt the ice into a pound of water at 32°F. If that pound of water is now heated to 212°F, 180 *Btu* is required (212 − 32). To completely boil the pound of water now will require 970 *Btu*. The water vapor produced will still be at 212°F. All total, 1320 *Btu* of energy is required to heat the ice (26), melt it (144), raise it to boiling (180) and vaporize it (970). Note that the 970 *Btu* required to completely boil the water is a *definition* just as 144 *Btu* to freeze or melt is. The only calculations we did produced 26 which was 52 times the *specific heat of ice* (0.5), and 180 which was the temperature that the pound of water was raised between freezing and boiling.

From -20°F to 32°F, sensible heat is at work. Likewise, raising the temperature of the water from 32°F to 212°F involved sensible heat. The melting of the ice involved latent heat, as did vaporization of the water. All refrigeration except thermo–electric units and other esoteric methods such as vortex tubes and magneto-caloric refrigerators work around the principles inherent in latent heat.

Suppose for a moment that we can get water to boil at a lower temperature. Will it still take 970 *Btu* to vaporize a pound of it? For the most part, yes. If we can get water to vaporize at 40°F instead of 212°F, then we can transfer heat from objects hotter than 40°F into the water, using the transferred heat to *boil* the water. In the process, the objects supplying the heat to boil the water will get colder, ultimately reaching 40°F.

Evaporation is the basic principle behind the ancient Egyptian wine coolers, the desert water bag and modern refrigeration. Evaporation requires heat, and if we can obtain evaporation at a low temperature, then the heat needed for the evaporation process can be extracted from objects which we wish to cool.

Heat can neither be destroyed nor created, but it can be moved from here to there. To fully understand how modern refrigeration works will require that we understand how heat is moved via vapors which are compressed to liquid, and the liquid then allowed to evaporate.

9.4.9 The Gas Laws and Their Application

Many of us have learned firsthand how long it takes to cook on the top of a high mountain. The reason is quite simple. At 10,000 feet, atmospheric pressure is only $10\Psi^4$ instead of the 14.7Ψ at sea level. Water boils at only $193°F$ at 10Ψ. Thus, by lowering pressure, we lower the temperature at which vaporization occurs. If we subjected water to a low enough vacuum, it could be made to evaporate at $32°F$. That is a low enough temperature to yield cooling.

If we were to take a pressure cooker full of steam at $212°F$, and keep the temperature the same but increase the pressure to 25Ψ absolute, then the steam would condense back to water. This is the basis behind compressor refrigeration ... pressurize a gas and cool it, turning it back to a liquid so that it can be evaporated. Evaporation requires heat and that heat must be given up by surrounding objects ... the food we want cold.

While we have used water so far in our presentation, water is not too well suited for use as a refrigerant, because it would have to be evaporated in a vaccum to achieve significant cooling effect. A better refrigerant is one that exists naturally as a gas, but one which can be cooled to a liquid state at a moderate pressure.

One of the oldest refrigerants is ammonia. Because of its toxic nature in large quantities, it has largely been displaced by Freon[5], of one type or another. There are several types of Freon, but the most widely used grade is designated by the number 12. Freon is a

---

[4]We use the Greek letter Ψ to represent pounds per square inch.
[5]Freon is a trademark of E i Dupont and Company.

trademarked product, so it has become standard to talk about R–12 instead of Freon.

For comparison purposes, we will provide the characteristics of both ammonia and R–12. While ammonia is not used in small refrigeration systems, comparison with R–12 will serve to illuminate the important characteristics of a refrigerant.

Ammonia is a natural occurring element, but readily absorbed in water. An English chemist, Joseph Priestly first isolated ammonia in 1772. Freon, known technically as difluorodichloromethane was developed in 1930 by a U.S. chemist, Thomas Midgley[6]. Midgley demonstrated the safety of Freon by inhaling it.

At atmospheric pressure, ammonia evaporates at a temperature of -28°F. The latent heat at that temperature is 589 *Btu* per pound. A pound of evaporating ammonia will absorb 589 *Btu*, enough to freeze about 4 pounds of water if the water were already at 32°F. In reality, if you were to open a gallon of ammonia to the atmosphere, 13% of it would evaporate to cool the remaining 87% to -28°F. Eighty-seven percent of 589 leaves only 512 *Btu* to use for cooling other substances.

To keep ammonia in a liquid state at 86°F requires a pressure of 169Ψ. As a liquid, the specific gravity of ammonia is 0.66, and as a vapor, 0.74. Specific gravity, you may recall from Chapter 2 is the weight per volume relative to water.

At atmospheric pressure, R–12 evaporates at a temperature of −21.7°F and has a latent heat value of about 71 *Btu* per pound. A pressure of about 108Ψ is required to keep it liquid at 86°F. The specific gravity of liquid R–12 is 1.44, and as a vapor is 5.2, versus that of ammonia (0.74).

The idea behind refrigeration is to evaporate as much refrigerant as is necessary to perform the required cooling. One pound of ammonia will yield 589 *Btu*. To get the same cooling effect will require about 8 pounds of R–12. Based on these facts alone, you might wonder how R–12 has come to be so widely used. First, it is non-toxic. And though ammonia has over 8 times the latent heat value as R–12, larger volumes of it are required because it is so much lighter. Note again the specific gravity of R–12 as a vapor (5.2). A given volume of R–12

---

[6]Midgley worked at Delco from 1916 with Kettering, the inventor of the electric starter, Chapter 2.

weighs 7 times as much as the same volume of ammonia vapor. As we will see shortly, larger volumes require larger compressors. R–12 is more suited to small refrigeration compressors such as those in an alternate energy system, because more of it can be circulated with a small compressor. Thus, the lower latent heat value of R–12 is compensated for by the fact that smaller compressors can move more vapors and attain the same net cooling effect.

9.5 Insulation

No matter the refrigerator technology that you ultimately choose, it will not work without insulation and plenty of it. Thermal conductivity was discussed earlier. Now we wish to apply the principles of thermal conductivity to the issue of insulation.

Energy is expensive in the alternate energy system, particularly that consumed by refrigeration. It only makes sense to reduce losses as much as possible. One way to do that is to insulate, insulate, insulate.

The idea behind insulation is to impede the flow of heat into the refrigerator. Any material that presents good thermal resistance is a candidate. Insulation must not absorb any moisture. As we know, water is a good conductor of heat, not an insulator. Of course insulation should not rot, be eaten by rats, weigh excessively, or be too costly, but most importantly, it must not conduct heat.

Insulation is rated by how many *Btu* of heat will leak through a square foot of the material if it is one inch thick. The less the better. The rating number is called the *k factor*. The lower the *k* factor, the better the insulation. Insulation *k* factor is discussed in greater detail shortly.

The best insulator is a vacuum, but creating a vacuum for a large volume is out of the question. Dry and motionless air has a *k* factor of 0.16, not as good as a vacuum, but still, dry air is a very good insulator. Holding air motionless is difficult at best. The job of an insulation then is to hold air as motionless as possible while not conducting any heat through the insulation material itself. In this regard, insulation is nothing more than a means to captivate air.

There is only one type of insulation suitable to the small refrigeration system. It is polyurethane foam. It has a k factor of 0.17, and is available as sheets or two part liquids. The two part liquid foams can be purchased at any good plastics outlet. They are easy to use, though caution is needed when filling restricted cavities.

Mixing the two parts causes a chemical reaction and the result is a foam of bubbles which expands into the available spaces around a refrigeration box. The bubbles captivate air, preventing its movement. As the foam cures, it becomes rigid, providing additional strength. The foam is readily shaped by knife or abrasive tools. The foam must be protected from wear and direct immersion in water, but is easily covered with fiberglass and resin. The foam comes in different densities, with 2 pound density best suited for insulating.

Mixing of the two liquid parts should be done immediately upon joining. Stir vigorously, watching the foam for a telltale *flash*. When the chemical reaction has progressed sufficiently, the mixture makes a rapid but temporary color change. This flash is the signal to pour the mixture into the cavity prepared for it. The foam expands quickly and with great force. Do not attempt to constrain the expanding foam. It can collapse even very heavily constructed boxes if restrained. After curing, excess can be trimmed.

The expanding foam does not fill corners and crevices completely unless small batches are mixed and directly poured into them. Start out with small batches and experiment before trying a large batch.

How much polyurethane insulation is necessary? From looking at the amount of insulation used for the refrigerator in the typical *RV* or boat one can get the impression that not much is required. Such is not the case. Many builders of recreational vehicles (boats included) realize that their products sell on superficial appearance of quality rather than deep level functionality. Applying adequate insulation to achieve high performance might detract from the appearance of roominess in the galley. Strange as it may sound, empty space sells better than an efficient refrigerator.

9.5.1   Insulation k Factor.

Figure 9.1 shows the way insulation k factor is defined. The k factor is defined as the Btu's which will leak through a one foot square piece, one inch thick in one hour if one degree temperature difference exists from inside to outside. The box shown in Figure 9.2 is used for an example calculation of an insulated refrigerator box. Dimensions of the box are given in feet.

As noted, the equation which describes the relationship between the parameters is:

$$Q = (A)(k)(dt)/T$$

Units of Q are Btu per hour. A is the total outside surface area of the box, while k is the insulation factor. dt is the temperature difference from inside to outside, and T is the insulation thickness in inches. With an ambient outside temperature of $°100F$, and an internal temperature of $°40F$, the box in Figure 9.2 will lose 144 Btu per hour. That is equivalent to a pound of ice melting every hour.

Table 9.2 shows the effect of insulation thickness. Tabulated is the amount of heat lost from the box of Figure 9.2 when it is subjected to different internal temperatures. The box has inside dimensions of 2 feet wide, 2 feet long, by 3 feet deep. The box is surrounded by 4 inches of insulation. The assumed ambient temperature is that of a cool summer day in Baja California, 100°F.

Let's consider the box with four inches of insulation. Recall that a pound of ice melting absorbs 144 Btu. Just to feed the insulation loss of the box, one pound of ice must melt every hour. That is 168 pounds of ice a week! We haven't even started to cool anything off yet, and haven't opened the box to put anything into it. Suppose we cool 2 gallons of water a day for drinking. The weight of 2 gallons is 16.7 pounds. Cooling it from 100°F to 40°F will require 1002 Btu total $((60)(16.7))$. Dividing 1002 by 24 hours in the day yields 42 Btu per hour required to cool the water. What this exercise shows is that four inches of the best insulation will waste about 3.5 times the useful energy of cooling two gallons of water, given the box dimensions and the temperatures stated. This fact should stand out ... most of the

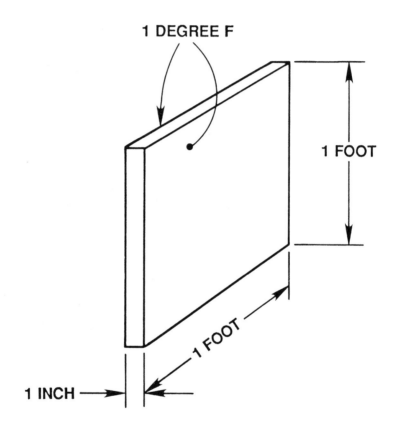

Figure 9.1: Insulation k Factor Definition

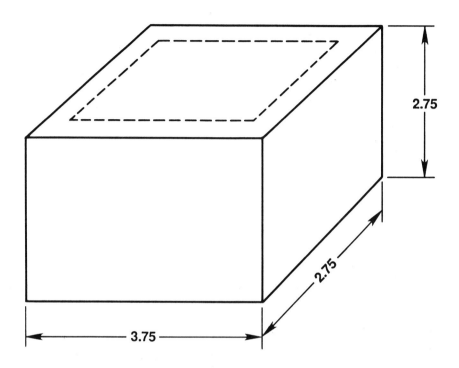

Figure 9.2: Insulated Box Calculations

| Insulation | (40 F) | (10 F) | (-20 F) |
Thickness, inches.	*Btu*/hour	*Btu*/hour	*Btu*/hour
1	575	863	1150
2	288	431	575
3	192	286	383
4	144	216	288
5	115	173	230
6	96	144	192

Table 9.2: Insulation Losses

energy that goes into refrigeration will be lost through the insulation unless you have an abnormally large in-use cooling need.

Not every box needs to have 12 cubic feet of internal space. And not every box needs to be designed for 100°F ambient. In a stationary application, (mountain mansion), space is not usually a premium so the insulation should be as thick as possible. The more insulation the larger the internal space can be for long-term storage of food for any given energy expenditure.

In a boat or *RV*, giving up extra volume for insulation means that some other features must be omitted. At the same time, boats and *RV*'s tend to migrate to warmer climates, increasing the need for insulation. Winter in Mexico may be 85°F and not 100°F, with temperatures cooler at night. This will relieve average demand. A smaller box will give up less heat through the insulation and is more appropriate for day-to-day use where forays to the market can be done fairly often.

Whatever the conditions, insulation pays dividends. Use as much as you can, and be prepared to generate enough energy for the insulation that you left out.

(Editors Note: We spent a summer in Baja cruising the Sea of Cortez. Despite our attempts to keep the icebox full of beer, it seemed to disappear at the same rate as the ice. This meant that our first duty ashore was to acquire both ice and beer. With a block under one arm, and a case under the other, we had what might be called a *Balanced Energy System.*)

9.5.2   Measuring Insulation Losses

Insulation losses for a refrigerator can be measured. While doing so you will also measure the losses due to door seals, and drains. Fill the box with a known amount of ice. Don't count on a block of ice to weigh 25 pounds ... weigh the ice with a scale. Turn off the compressor and measure the time that it takes the ice to melt.

Suppose that you can fit 100 pounds of ice into the box and it takes 50 hours to melt. That amounts to 2 pounds an hour or 288 *Btu* per hour. That loss is the particular loss that occurs for the relative temperatures involved in your experiment. You should periodically measure the inside temperature as well as the outside ambient, taking the average temperatures for your calculations.

Note that in the heat conduction equation, temperature difference is a multiplying parameter. This means that you can calculate the loss for other temperature differences. Suppose that your test case above was done with an average inside temperature of $40°F$, and an average outside temperature of $80°F$. Thus, dt, the temperature difference, was $40°F$. What is the loss if the ambient temperature is $100°F$? Since the difference is now $60°F$, (or 1.5 times 40) , then multiply your test results by 1.5. Where the test yielded 288 *Btu*/hour at $80°F$, then at $100°F$, the insulation will lose 432 *Btu* per hour. Such a loss would require many Amp-hours to produce, and would likely be impossible without long hours of engine charging.

Insulate !

Chapter 10

Compressor Refrigeration

10.1   General Information

The basic compressor refrigerator is shown in Figure 10.1. Let's start
with the compressor. It is usually a piston driven apparatus much
like an automotive engine. Small compressors will have a single cylinder. The intake to the compressor is low pressure vapor returning
from the evaporator. Vapor is compressed to the proper pressure
and discharged to the condenser. The condenser is a series of tubes
which are cooled by air or water. Because the vapors are cooled in
the condenser, but still under pressure, they condense into a liquid.
The liquid is stored in a tank, often called a receiver. Sometimes the
receiver tank is part of the condenser.

   The liquid in the receiver is still under pressure from the compressed vapor. This forces the liquid to flow to the expansion valve.
An expansion valve is just a metering device . . . that is, it allows a
controlled amount of liquid to pass into the evaporator. Once inside
the evaporator the liquid is no longer under pressure, and begins to
evaporate. As it evaporates it absorbs heat from its surroundings,
the evaporator. The evaporator in turn must conduct heat from the
refrigerator box. All the liquid must be completely evaporated by the
time it gets back to the compressor because liquid does not compress
. . . compressor destruction results if liquid returns to it.

   To help prevent compressor damage, expansion valves are usually controlled by sensing the temperature at the evaporator. If the

compressor end of the evaporator gets too cold from liquid which is still evaporating, then the valve is throttled back to release less liquid into the evaporator. With less liquid to evaporate, the suction end of the evaporator will run *starved*, and thus, warm up. Warming at the suction end will cause the expansion valve to pass more liquid. To make maximum use of the evaporator, it is necessary to allow as much liquid to pass as possible without flooding the return line to the compressor.

Such is the refrigeration cycle when all goes as planned. We need to examine each of the components in the system in greater detail to see what the limits of performance are. The bottom line is the overall efficiency of the system. How much energy must be input to the refrigerator to get the desired cooling effect? The answer to this question lies in the individual efficiencies of the compressor, condenser, evaporator and the motor driving the compressor. In general, for every *Btu* of energy required to drive the compressor, 1–3 *Btu* of cooling are produced. At first this might appear to be creation of energy ... more *Btu*'s out than in. Actually, energy is not created, rather the compressor, condenser and evaporator are tools to move heat from one place to another. Heat is moved from the evaporator to the condenser. Compressor refrigeration does not generate energy any more than a water pump creates water.

Before we present data regarding the individual parts of the system, a reference is needed. The reference points are used to define a standard set of operating conditions. Under standard conditions, the evaporator temperature is $5°F$, and the condenser is $86°F$. These operating conditions are hard to achieve with inefficient components ... the reason that many refrigeration installations are less than successful.

Refrigeration systems rarely work at these standard conditions. Condensers in particular often run much hotter than $86°F$. In hot places, like Puerto Escondido, Mexico, water temperatures can exceed $86°F$ so even a water cooled condenser will run hot. Air cooled condensers in warm climates are hard pressed to maintain $10–15°F$ above ambient, even with lots of air circulation.

Evaporators, of course do not all run at $5°F$. Freezers holding ice cream for long periods should be maintained about $-10°F$. The evap-

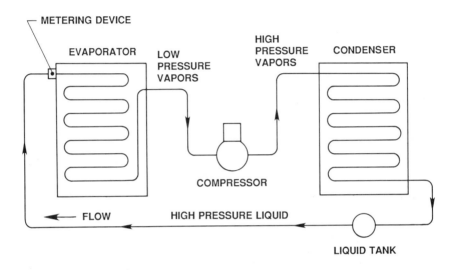

Figure 10.1: Basic Compressor Driven Refrigeration System

orator may be running at -35°F to -20°F to achieve this. Conditions in a small boat are quite different than shoreside ... efficiency and performance are issues which must be considered early in the design phase.

The evaporator shown in Figure 10.1 may be more than the simple flat plate evaporator that we are familiar with. Often, the evaporator is part of a holdover plate assembly which stores cold energy in the same manner as ice. The holdover plate system is described in greater detail later.

10.2   Compressors

Compressors supply the force that keeps the system operating. Figure 10.2 shows a *hermetically* sealed compressor as well as an *open* unit. The open type compressor on the left requires a motor to turn it. The hermetic compressor on the right has a motor hermetically sealed with the compressor. The open type compressor is more efficient than the hermetic type, but requires effective seals around the crankshaft.

The compressor takes vapor from the evaporator and applies the pressure which allows it to be condensed to a liquid. Small compressors are always piston operated, with *suction* and *discharge* valves. Valves in a compressor are not push rod operated as they are in an engine, rather they are reed or *flapper* valves which are thin metal plates. The piston is connected to a crankshaft, which, when rotated causes the piston to traverse the cyclinder. Automotive air conditioning compressors do not use a crankshaft, but instead employ a *swash* plate.

The parameters which affect the efficiency of the compressor are the suction valve spring pressure, clearance between the piston and the head, and the leakage past the piston rings. The rings in particular wear and eventually lead to a loss in refrigeration. It should be noted that small compressors may not have rings, but rely on oil grooved pistons.

Larger refrigeration systems are fully instrumented so that efficiency and preformance can be monitored. Keeping the compressor alive is the primary concern. Suction pressure, discharge pressure,

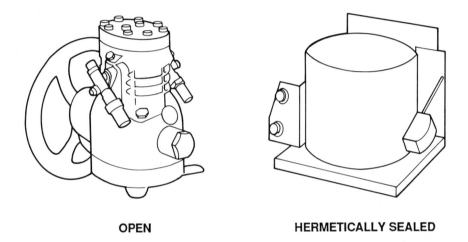

OPEN   HERMETICALLY SEALED

Figure 10.2: Compressor Types

evaporator temperature and condenser temperature are all measured. If you build your own system, you should consider this same level of instrumentation. Given the high cost of refrigeration equipment and its cost of operation, it will pay to achieve optimum performance. As we have noted in the past, measurement is the cornerstone of performance, and pressure transducers are relatively inexpensive.

Though internal construction details vary from one compressor to the next, there are really only two kinds which concern us ... the open type and the hermetically sealed type. The open type compressor is one which is driven by a motor mounted external to the compressor. The hermetically sealed compressor has a motor which is sealed in the same enclosure as the compressor itself. There are advantages and disadvantages to both.

Under high temperature refrigerator conditions, both the input or low pressure side and the output or high pressure side of the compressor operate at a pressure greater than atmospheric. For low temperatures associated with freezers and holdover plate systems, the low pressure suction line operates below atmospheric pressures.

Seals in the compressor are required which prevent leakage of refrigerant to the atmosphere. By sealing the motor with the compressor, seals are required only on the whole assembly. With the motor driven electrically, the assembly of motor/compressor can be sealed by simply sealing around the wires which exit. Hermetically sealed units also run quieter than do open type compressors. Note that domestic refrigerators all use hermetically sealed compressor/motor assemblies, as do most of the small *DC* refrigerators sold for marine service.

The open compressor, on the other hand, must have very effective seals around the crankshaft. These seals will eventually wear out, and therefore, service on an open compressor may be required more often. When seals are not exercised frequently they tend to dry up and leak. This is the primary reason that car air conditioners need service so often. Even during winter, owners should run their air conditioners at least once a week to keep the seals coated with oil and thus functional. The same is even more true of engine driven refrigeration systems because the compressor operates at low suction pressures, below atmospheric pressure for freezers and holdover plates. The automotive air conditioning compressor was not really designed for evaporators less than freezing, so expect more seal problems when used as a compressor for marine refrigeration. Other open type compressors have been designed for cold work, but cannot tolerate the wide *RPM* range of engine drive.

Despite a problem with seals, when not run frequently, the open type compressor has a distinct advantage over the hermetically sealed unit. The motor which drives the hermetically sealed unit produces much heat. This heat is conducted into the refrigerant itself and must be extracted by the condenser. The condenser will have to be bigger, and more energy will be expended to keep it cool. Naturally the overall efficiency of the system is reduced. Since we are after maximum efficiency, the open type compressor must be considered. In a typical domestic refrigerator connected to infinite *AC* power, efficiency is not an issue and the hermetically sealed motor/compressor is the appropriate choice. For very small alternate energy systems, the ease of installation and lack of maintenance make the hermetically sealed motor/compressor a good choice. Larger alternate energy systems, for reasons of efficiency should use an open compressor belted to a separate motor.

The open type compressor has yet another advantage. A typical failure of a hermetically sealed compressor/motor is burnout of the motor. Motor cooling relies on enough suction vapor for its cooling. A blockage in the system, or failure of an expansion valve can reduce vapor flow to the point that the motor burns out. The whole assembly must be scrapped, and the remainder of the system thoroughly cleaned before a new compressor is attached. After a burnout, a suction line filter is recommended just ahead of the compressor to catch any remaining residue.

Compressors are oiled by several methods. The simplest method is the splash system where the piston rod and crankshaft splash the oil on each revolution. Larger compressors have an integral oil pump. Hermetically sealed units usually use force feed oil pumps. The automotive compressor relies on the swash plate rotating through the oil sump. Because of differential pressures inherent in the swash plate compressor, quite a bit of oil circulates with the refrigerant. Extra oil is often needed in holdover systems because of accumulation in the condenser or evaporator. Note that circulating oil decreases the effectiveness of both condenser and evaporator. For highest efficiency, an open style compressor with splash lubrication is best. Such a compressor should be run as slow as permissible and still achieve the refrigerant capacity/volume required. The type of oil used in compressors is critical. Never use any oil except the type specified by the compressor manufacturer.

Compressors are rated in several ways ... *Btu* per hour, horsepower required to drive them or simply *bore and stroke*. The rating is important to us, but before getting into those details let's first consider important parameters of condensers and evaporators.

10.3   Condensers

Figure 10.3 shows both air cooled and water cooled condensers. On the left is an air cooled unit which has fins around the tubes which carry the refrigerant. On the right is a water cooled unit which submerges the refrigerant tubes in a tank of water. The center condenser is a *tube in a tube* type. One tube conducts water, while the other tube carrys the refrigerant. This type of condenser is most efficient.

A condenser must accept the hot vapors given off by the compressor and cool those vapors to a liquid. To cool the hot vapor, the condenser must conduct heat out of and away from itself. Earlier we discussed the idea of heat conduction and showed that it depended on the surface area and the type of material conducting the heat. One of the best heat conductors is copper, so we would expect that the better refrigeration systems use copper condensers, particularly air cooled systems.

The condenser must remove all the heat picked up by the evaporator plus the heat gained by compression, and the heat produced by the motor driving the compressor, if the compressor is hermetically sealed. The heat that the condenser must conduct away will usually be 1.2–1.5 times the heat conducted through the evaporator. To maximize efficiency, a condenser with good thermal conduction to the external environment is necessary.

Condenser efficiency is directly related to surface area. The more surface area there is to conduct heat, the more efficient it is. A con-

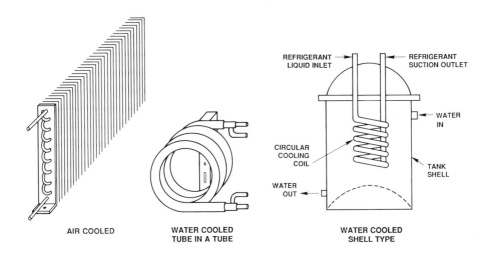

REFRIGERANT
LIQUID INLET

REFRIGERANT
SUCTION OUTLET

WATER
IN

CIRCULAR
COOLING
COIL

TANK
SHELL

WATER
OUT

AIR COOLED

WATER COOLED
TUBE IN A TUBE

WATER COOLED
SHELL TYPE

Figure 10.3: Condenser Types

denser is also required to have a large enough volume so that the compressor does not have to labor under a great amount of back pressure. Coming into the condenser are high pressure gas vapors. Leaving, there should be nothing but liquid refrigerant. The pressure on the liquid will be essentially the same as the compressor discharge pressure, although some losses will be incurred through the condenser and the liquid line filter/drier.

Condensers may be air cooled or water cooled. A refrigeration system using R–12 will typically operate the condenser at $90°F$ and $115Ψ$ pressure. To keep a condenser at $90°F$ will likely require a lot of air flow. With air temperatures over $90°F$ the battle is lost. (Condensers in home refrigeration often run about $125–130°F$, but we seek much higher efficiency.)

Water cooling is more efficient than air because the available water is usually less than $90°F$, and thermal conductivity of water is so much higher than air. (Remember our prior example of hand in the oven versus hand in boiling water.) The cooler the water, the less water necessary to cool the condenser. Since the water will likely have to be pumped, the energy required for pumping must be included in the energy cost of refrigeration. Only enough water to remove condenser heat should be pumped for maximum efficiency. To control water flow, a centrifugal pump can be used with a throttling valve. (More on this later.)

Condensers are constructed from a length of tubing which is usually doubled back on itself several times to provide a great length in a small space. If air cooled, then the tubes have metal fins attached which provide the required surface area for cooling. Thick cooling fins are necessary if maximum heat transfer efficiency is desired ... thin fins are less expensive to make, and most often used.

When a condenser is water cooled, the tubes which carry refrigerant can be immersed in a container of water. Normally the water must be exchanged to achieve the desired cooling effect. Water exchange may be done on a continuous basis or may be done intermittently. In small refrigeration systems, continuous water exchange may not be practical unless you live next to a small stream that provides gravity fed water. The amount of water required for condenser cooling is on the order of several gallons an hour for small sytems to several gallons per minute for larger ones. Small pumps that produce several gallons an hour are not generally very efficient and may not be self-priming. In a boat that occasionally heels, loss of prime may mean loss of refrigeration. A more practical solution is to use intermittent water exchange. This method periodically exchanges the water in the condenser by pumping in a fresh load. The pump then shuts down until a new load of water is required. The pump may be activated by a timer or by a temperature sensor on the condenser. This latter method is more expensive but also more effective since it only pumps water when the condenser begins to get warm.

Large condensers such as those required for rapid pumpdown of holdover plates will require 1–4 gallons of water per minute, depending on the system capacity rating. Larger systems should use the tube in a tube condenser which has the highest efficiency rating. A tube/tube condensers is made with one tube inside the other. The inside tube carries the water while the outside tube carries the refrigerant. For any water other than pure water, the water carrying tube must be made from cupronickel which will not corrode from galvanic action.

While it is quite simple to fabricate your own tube/tube condenser, there are units available commercially at a reasonable price which are made with convolutions in the inside water tube. The convolutions create turbulence in both the water and refrigerant. An increased surface area is also achieved via the convolutions. The combination of turbulence and increased surface area add to efficiency ... the result is less energy spent pumping water.

Whether condensers are air cooled or water cooled, they must be clean to operate at peak efficiency. Air cooled condensers tend to accumulate a layer of greasy dust which insulates them. Water cooled condensers, though cupronickel and quite anti-fouling, acquire a layer of slime, particularly in warm saltwater conditions. Spiral or helical wound tube/tube condensers are harder to clean than the straight tube variety, some of which have easily removable headers to allow cleaning. Cleaning the condenser should be a regular maintenance concern. All too often manufacturers of refrigeration equipment do not provide the ready access necessary to clean the condenser. Once a refrigeration system is sold it is easy to blame its failure on others ... poor installation, not enough batteries, not enough insulation, too

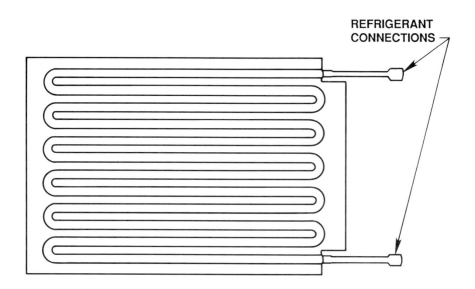

Figure 10.4: Flat Plate Evaporator

big a box. All these conditions are true in too many cases.

10.4   Evaporators

The evaporator is so named because the refrigerant evaporates inside the apparatus. Evaporation absorbs heat from its surroundings which provides the desired cooling. In principle, an evaporator can be constructed just like a condenser since both are required to conduct heat to or from their surroundings. In the small refrigeration system however, the evaporator does not usually have fins around the tubes like the condenser. Instead the tubes are formed into flat aluminum plates. Aluminum is a good heat conductor as well as being light and corrosion resistant.

Figure 10.4 shows a typical flat plate evaporator. The tubes shown are usually formed in the aluminum plate itself, rather than being separate tubes. This type of evaporator is used in a continuous cycle refrigeration system, and only cools when refrigerant is evaporating inside.

The large surface area of the plate conducts heat readily. Natural convection is used when the surface area of the evaporator is large enough. Cold air falls while warmer air rises, setting up some circulation of air around the evaporator. By placing baffles around the evaporator, better circulation can be achieved and a cooler area can be obtained for some foods.

Figure 10.5 is an evaporator which contains a *eutectic holdover* solution used like ice. The tubes shown are internal to the holdover tank. The tank contains a brine solution which is frozen solid whenever the holdover plate is *pumped down* by running the compressor. When the compressor is not run, the holdover plate absorbs heat just like a block of ice, but usually at a lower temperature.

An evaporator inlet is usually at the top. Refrigerant is liquid when it enters and flows toward the bottom. As it flows, it evaporates, cooling the evaporator as well as the remaining liquid refrigerant. Earlier we mentioned that 13% of ammonia was used just to cool the remaining liquid. For R–12 evaporating at 5°F from a liquid at 86°F, about 27% of the R–12 is used just to cool the remaining R–12. The remainder, 73%, is available to cool the surrounding box. Later, the calculations necessary to determine refrigerating effect are presented.

An evaporator must have sufficient surface area to conduct the required *Btu*'s from the box. It must also have a long enough length of internal tubing so that the liquid entering the top is completely evaporated by the time it reaches the outlet on the bottom. Surface area and evaporation length go hand in hand. A third requirement of the evaporator is adequate circulation through it. Any restriction in circulation will result in back pressure on the compressor with a reduction in compressor efficiency.

To increase circulation through the evaporator a technique called *headering* is used. The liquid entering the evaporator is split into several tubes by a *header*. With the several tubes conducting refrigerant in parallel, circulation is improved and with it, overall efficiency. Evaporators in most small systems are not headered since designers are more concerned about cost than ultimate performance. For holdover systems with several plates, headering can be effective by

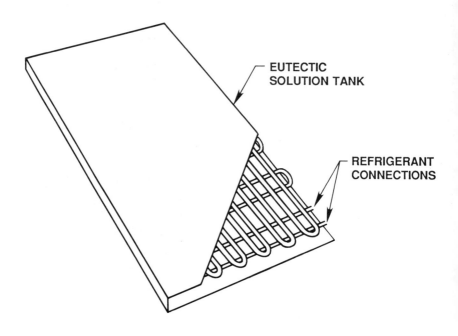

Figure 10.5: Holdover Plate Evaporator

simply driving the plates in parallel.

It would seem that circulation could be improved just by going to bigger tubing inside the condenser. In general, however, there is a balance between tubing size and the rate of flow of refrigerant through the tubes. If the velocity is too slow, then oil which is dissolved in the refrigerant will cling to the inner surface area instead of being returned to the compressor. Oil is an insulator, so the effectiveness of the evaporator is reduced. Too much oil in the evaporator may deplete the oil in the compressor. More details about refrigerant velocity are presented later.

10.5 Liquid Metering Techniques

In Figure 10.1 an expansion valve is shown. This is one of two methods used to control the amount of liquid entering the evaporator. The other method employs a capillary tube. The purpose of both methods is to control the amount of liquid entering so that all the liquid is evaporated by the time it gets to the evaporator outlet. As noted earlier, no fluid must be allowed to return to the compressor.

The capillary tube operates on the simple idea of restriction. Liquid refrigerant is passed through the capillary tube which is several feet of very small tubing. The tube has an orifice of 0.03 to 0.06 inches in diameter. The length of the capillary tube and its diameter are fixed by the designer of the system. To be on the safe side, the compressor is operated at slightly more back pressure than necessary by restricting the flow through the capillary tube. This also means that a portion of the evaporator is not being used. The capillary tube is not used for operating efficiency. Since it has no moving parts and is just a tube, cost is minimal. The capillary tube does have a benefit, however. When the compressor cycles off, pressure throughout the system equalizes. When the compressor restarts, it isn't required to start under a load. Such a feature is important if the compressor is driven by an *AC* motor with low starting torque.

Figure 10.6 shows the connections to and from the expansion valve, including the temperature sensing bulb, and pressure equalization tube. The pressure from the sensing bulb opens and closes the valve

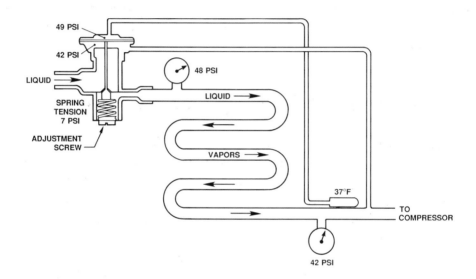

Figure 10.6: Expansion Valve Circuit

to maintain a constant temperature at the bulb.

The expansion valve shown in Figure 10.6, is obviously more complex than the capillary tube ... likewise it is more efficient, and the system will perform properly over a wider range of conditions. There are different kinds of expansion valves, but the best one to use senses the temperature at the evaporator outlet and adjusts the inlet flow so that the outlet contains no liquid. Other types of valves work strictly on pressure sensing, and are not as easy to apply properly. With the temperature sensing valve, any liquid reaching the outlet will be evaporating and cools the outlet temperature sensor. The colder temperature acts on the expansion valve to reduce the flow of refrigerant into the evaporator.

The sensing bulb of the expansion valve is filled with a gas which expands under temperature. This causes the pressure to increase. The bulb pressure is applied to one side of the expansion valve. Opposing the pressure from the sensing bulb is a spring and the pressure of the circulating refrigerant inside the valve. For proper operation, the pressure at the output of the evaporator where the sensing bulb attaches must be the same as the pressure at the valve. If a large pressure differential exists between the valve and the output of the evaporator, an external pressure compensating tube must be used. Holdover plates which operate at subzero temperatures can cause problems if used without the external equalizer type of expansion valve since less pressure differential can be tolerated at colder temperature. More specific details are presented later.

10.6 Other Refrigeration Components

So far we have presented compressors, condensers, evaporators and metering devices. In addition to these parts, other components are used to complete a system. Besides the tubing which is necessary to connect the above parts, other items are worthy of discussion ... filters, driers, a refrigerant sight glass, and pressure sensing switches and transducers.

Outside Diameter	Inside Diameter	Wall Thickness	External Sq. in.
1/4	0.190	0.030	0.0491
3/8	0.311	0.032	0.110
1/2	0.431	0.032	0.196
5/8	0.555	0.035	0.307
3/4	0.680	0.035	0.442

Table 10.1: Parameters for Soft Copper Tubing

10.6.1 Refrigeration Tubing

As noted, there is both high pressure and low pressure in the system. Using R–12, the high pressure side is operating around 115 Ψ. The low pressure side is operating slightly higher than atmospheric pressure for refrigerators, and sometimes less than atmospheric pressure for freezers. Usually steel tubing is used for both the high pressure side and low pressure side in domestic refrigerators. In a small stationary alternate energy system, copper is a better choice. Copper can also be used in mobile systems as long as vibration is prevented. Parameters of soft copper refrigeration tubing is given in Table 10.1.

For a system which drives a compressor from an engine, a flexible high pressure tubing such as double braided refrigeration hose is necessary from the compressor to provide vibration resistance. R–12 can escape normal hydraulic type hose ... use only authentic refrigeration hose.

10.6.2 Liquid Receiver

Another part in the system is the liquid receiver. This may be a special tank which is connected between the condenser and the evaporator, or it may be built into the condenser. The purpose of the receiver is to accumulate liquid that is not currently needed in the evaporator. When a heavy load is placed on the evaporator the level in the receiver will fall as the expansion valve is opened to increase flow. Later as the load abates, the receiver will begin filling again with liquid.

10.6.3 Filter/Drier

Moisture in a system using R–12 will cause havoc. Parts will corrode, expansion valves will form ice blockage, and the wiring in a hermetically sealed compressor can be damaged. To prevent these problems a *drier* is recommended. A drier is nothing more than a cartridge with a moisture absorbing chemical inside. Silica gel, activated alumina or calcium sulphate are used. The drier is usually placed in the high pressure liquid line since its volume can be smaller there than in the low pressure vapor line. The drier is placed between the condenser and the evaporator, just as is the receiver tank.

Filter/driers must be sized for the rating of the system. No harm is done by selecting a unit with greater capacity than needed, since the pressure loss will be less on a larger unit. The liquid line filter/drier should be installed after the condenser, but just before the refrigerant sight glass.

Sometimes a filter/drier is used in the suction line, just preceding the compressor. Its purpose is compressor protection from residue left in the system during construction. Despite attempts to keep all foreign matter out of the system, many small particles may be inside, and they can quickly destroy a compressor by damaging valves or gouging cyclinder walls. As noted earlier, anytime a compressor is replaced in a system after a burnout, it is good practice to add a filter/drier ahead of the compressor. A filter/drier in the suction line will require a larger rating than one in the liquid line.

10.7 Heat Exchanger/Accumulator

A combination heat exchanger and accumulator should be standard on any holdover plate system. Refer to Figure 10.7. The hot liquid is circulated through the cool vapors returning to the compressor. Any liquid leaving the evaporator is trapped in the accumulator until it evaporates. The accumulator protects the compressor, while the heat exchanger increases efficiency. The liquid is cooled below the con-

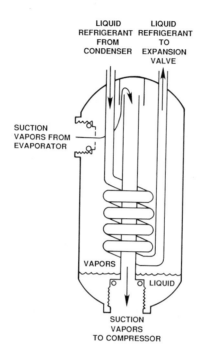

Figure 10.7: Heat Exchanger/Accumulator

densing temperature. This subcooling of the liquid guarantees that it won't be evaporating prior to passing the expansion valve. If any liquid evaporates ahead of the expansion valve, not only is the refrigerating effect wasted, but the valve capacity is seriously degraded. Excessive pressure losses in the liquid line can lead to such *flash gas* ahead of the valve.

The vapors leaving the evaporator are fed to the exchanger. The exit from the exchanger is at the top, thus any liquid refrigerant or oil trapped in the evaporator is accumulated. In particular, the accumulator part of the heat exchanger/accumulator plays an important role when the compressor is first started. With the system lying idle, liquid refrigerant and oil drains to the bottom of the evaporator. Without the accumulator, there would be a slug of liquid returning to the compressor when it was first started. As the system runs, the accumulated oil and refrigerant evaporates and returns to the compressor as vapor.

Because some compressor oil is always circulating with the refrigerant the oil level of the compressor must be checked after the system has been run. Generally in a holdover system it is a good idea to fill the compressor from 3/4 full to nearly full before the compressor is first turned on. Then after a full pumpdown cycle verify that the oil level is 1/2 to 3/4 of the compressor sight glass.

Note that most heat exchanger/accumulators have a service valve in the vapor side. This is often used to measure the pressure at the evaporator outlet when adjustments to the expansion valve are made. As mentioned, this is also a ready spot to mount an electronic pressure transducer.

10.7.1 Liquid Sight Glass

A liquid sight glass should be installed just ahead of the expansion valve. The sight glass should not have any bubbles when the system is operating normally. Glasses are available which also turn colors to indicate the amount of water in the system. As mentioned, water in the R–12 has disastrous effects.

10.7.2 Pressure Limit Switches/Transducers

A small hermetically sealed DC motor/compressor system does not usually require low pressure or high pressure limit switches. Such switches are essential on larger systems, particularly on holdover plate designs. The purpose of the switches is to interrupt the power to the compressor if the pressure limits are exceeded. The low pressure switch must be adjusted to the pressure which represents freezing of the eutectic brine in the holdover plate. High pressure switches can be preset at a pressure of about 200–250 Ψ.

You may choose to instrument your system with an electronic pressure transducer in the low side and a normal pressure switch on the condenser side. Not only will the transducer assist in the adjustment of the expansion valve, but the signal from it can also be used to trip a cut-out relay which removes power from the compressor.

10.8   Continuous Cycle Refrigeration

Domestic refrigeration operates in a continuous cycle. That is to say, the compressor is turned on when the refrigerator warms up and is turned off when the box is sufficiently cooled. The system does not store *cold*, other than that which is stored in the cooled foodstuff. The compressor may never turn off if the box cannot be pumped down to the desired temperature.

To operate in this continuous cycle, energy must be available to power the compressor whenever temperature in the box gets too warm. When tied up to the power pole, energy is available full-time except for rare occurrences of power failure. Such is not the case for alternate energy systems. To operate in the continuous cycle within the alternate energy system requires that batteries be present and sufficiently charged. Because refrigeration is a large energy consumer, the batteries must be appropriately sized. This in turn means that charge sources must be larger and/or operate for longer hours. The cost to buy, operate and maintain larger electrical systems must be accounted for in the true cost of refrigeration. In virtually all cases, the addition of refrigeration demands an upgrade in the electrical system. Without refrigeration, an electrical system with 30 to 50 Ah per day capacity may be adequate. If you add refrigeration, plan on upgrading the electrical system to support 100 to 200 Ah per day.

Small *DC* operated compressors are popular. They are relatively simple to install, and even simpler to operate. As long as the batteries can supply power, the refrigerator can cycle at the demand of its thermostat. In practice these small *DC* systems don't operate up to the expectations of their owners. All too often the refrigeration salesperson did not explain the need for a performance electrical system, or the owner was unwilling to invest in the proper power system to drive the refrigerator.

Even poorly balanced systems will provide some degree of utility in the northern climates. When these systems visit warmer weather both the electrical system and the refrigeration system show their true colors ... red hot. We devote a chapter to the *balanced energy system* and won't cover the same material here, but do emphasize that a continuous cycle refrigeration system is much more expensive to own and operate than the price tag on the refrigerator indicates.

10.9 Intermittent Cycle Refrigeration

The intermittent cycle system is one which does not operate at the demand of a thermostat but rather relies on a *holdover* plate to store a frozen brine. The brine melts between the periods when the compressor is not operating. In a sense, the holdover tank can be regarded as a battery which stores *cold* instead of electrical energy.

The intermittent cycle or holdover system is often more suited for use in the alternate energy system than is the continuous system. The continuous system requires a continuous source of energy. Storing energy in batteries is an expensive proposition, given the limited number of discharge cycles which are available.

A holdover plate has no intrinsic number of cycles which it can be operated. Used properly, a holdover system will perform for a long time with minimal maintenance. Such a system does require a large amount of energy at one time ... when the compressor runs, the holdover plate must be frozen solid. The latent energy of melting is recovered slowly from the holdover plate when the compressor is not running. Except for very low energy systems, (without refrigeration) or systems with many solar panels, some sort of fuel operated generator must be used. To pump down the holdover plate rapidly, it is desirable to run the generator only when all of its available energy can be consumed or stored. In this situation the holdover refrigeration system has great merit. During the time the generator is run to charge batteries or perform other periodic work, the holdover plate is frozen. (With both a large alternator and a refrigeration compressor operating from a single engine, plan on 8–16 horsepower consumed.)

A compressor in a holdover system must be much larger than one in a continuous system. All the refrigeration for a full day must be generated in a short period, say 45 minutes to an hour. The load on the compressor will be higher than might be expected. The pressure differential from input to output is higher because of the greater evaporation taking place, so the compressor has to work harder. Also the evaporator temperature is colder, translating into more compressor load. To achieve a rapid pumpdown, the evaporator temperature must be much colder than the temperature at which the plate freezes.

The length of time that a holdover plate will keep the box cold is naturally dependent on the cooling required by the box and its contents. The amount of stored energy in a holdover plate is directly related to the weight of the internal solution and its specific latent heat value during melting. More about these details is presented later.

The holdover plate serves a dual purpose in the system. Besides storing energy, the holdover plate also performs as an evaporator. Refrigerant is pumped through tubes which run inside the plate. Evaporation which occurs inside the tubes cools the tubes and the brine solution inside the holdover plate. High quality holdover plates use tubes with attached fins or other metal structural parts which distribute the heat more effectively and thus permit faster pump down. Holdover plates are made in various sizes but all contain tubes for circulating refrigerant as well as a chemical solution which stores energy of latent heat. The solution, called a *brine*, is a combination of water and other chemicals.

The surface area of the plate must be large enough to extract the heat from the refrigeration box. At the same time the volume in the plate must be sufficient to retain enough energy to last between pump downs. The design of a holdover plate should balance the surface area and volume requirements. For long times between pump downs, a large volume is necessary. A correspondingly large surface area is not needed, unless the plate is being used in a large box. The fact is, that if you provide enough brine capacity, you will always have more surface area than needed to keep the box cold.

Another important parameter of the holdover plate is the rate at which the plate can be frozen. This is analogous to how fast a battery can be charged. As it turns out, the factor which limits the speed at which a plate can be frozen is the amount of external surface area of the tubes which are internal to the plate. The automotive air conditioning compressor is of much greater capacity than the absorption capacity of the holding plates.

A holdover plate affects the compressor by presenting lower suction pressures. Once the holdover plate is frozen, the expansion valve will throttle back the amount of refrigerant which enters. This will increase back pressure on the compressor and also decrease suction pressure even more as the temperature continues to decrease. Eventually, if the compressor is not shut off, the crankshaft seals will give way, expelling refrigerant and opening the system to air. Compressors in holdover plate systems must be protected against low suction pressure as well as high discharge pressures. The most positive way to provide protection is to sense the high side and low side pressures and shut off the compressor when either pressure exceeds its limit. Some users elect to put the compressor on a timer which is set to operate before the plate reaches final temperature. This won't always protect the compressor . . . what happens when the plate is already frozen and the compressor is started?

Many holding plate systems are not as effective as they might be, simply because they have not been completely frozen. This is the result of using a compressor with much greater capacity than the absorption capacity of the holdover plate. On the other hand, a high rate of compressor failure can be attributed to operating at too low a suction pressure. Too low a suction pressure is the result of running the compressor too long after the plate is completely frozen, or once again, using a compressor too large for the plate. The most efficient and reliable system will be one that has a compressor matched to the absorption rate of the holdover plate, and a low pressure cutoff switch set properly.

We favor the intermittent system because it does not require lots of expensive batteries to operate properly. The intermittent system is not as self tending as the continuous cycle system which draws from the batteries as it needs . . . that is the usual assumption. Whether this is true or not depends on how the system was designed. If you choose a compressor belted to an engine, say your propulsion engine, then running that engine will be a periodic chore. In warmer weather such as that found in Mexico, refrigeration may require engine operation twice or thrice daily if the holdover plates are too small for the imposed load.

An intermittent system with a DC operated compressor, rather than an engine driven compressor has many merits. Such a compressor is normally only operated when the alternator is charging. A large amount of current is required at that time to pump down the box, rather than a small but continuous current of a *DC* refrigerator without a holdover plate. The intermittent *DC* system can also be run from batteries which are recharged at a slower rate, say with solar panels or wind generators. Therefore, refrigeration can be maintained without daily running of the engine if sufficient electrical energy is available. This allows the system to run unattended for short vacations, assuming that the system can be started automatically and shut off again when the holdover plate is pumped down. A *DC* operated intermittent compressor system will likely be more closely matched to the absorption rate of the holdover plate so that even though the compressor is rated much less than the engine driven compressor, total run time will not be proportionately longer.

A mixed system is another option. Such a system has an engine driven compressor which supplies a quick pump down, and a *DC* driven compressor which operates when the holdover system has mostly melted. This type of system can be quite effective in sunny climes. Batteries can be topped off in the evening in preparation for use of electrical lighting. At the same time, the holdover system is pumped down. During the following day, solar panels can partially or wholly drive the *DC* refrigerator which is beginning to take up where the holdover plates left off. A system balanced in this manner can operate efficiently with a single engine run per day.

The key to any holdover system is the amount of eutectic solution held by the plates. The volume and the time between pumpdowns go hand in hand.

10.9.1   Holdover Brines

The efficiency of a holdover system is very much affected by the type of solution inside the plate. In particular, its specific latent heat value should be high enough so that a small volume will be effective. Brines are a solution of water and other chemicals which decrease the temperature at which the solution freezes. Salt brines can be used, or other mixtures which are explained later.

As the temperature of a brine goes below $32°F$, water crystals form. This increases the density of the remaining solution. At some point the salts in the brine are not soluble in the remaining water.

Freezing Temperature Fahrenheit	Percent Calcium Chloride	Specific Gravity	*Btu* per Gallon
25.5	8	1.07	1136
12.5	15	1.13	1079
6.5	17	1.15	1063
-5.5	21	1.19	1042
-15.5	23	1.21	1031
-25	25	1.23	1023

Table 10.2: Calcium Chloride Brine

The solution freezes solid at this point, which is called the eutectic temperature.

Because the salts can precipatate to the bottom of the tank as the brine is frozen, an immobilizing gel is used. If the salts are allowed to fall to the bottom, they will not dissolve readily back into the water when it thaws. This would mean that the brine would freeze at a much higher temperature on subsequent cycles. Other brine additives reduce the amount of expansion that occur on freezing.

Holdover plates made for commercial use, such as in refrigerated delivery vehicles use a calcium chloride or sodium chloride brine. A small amount of lime is added to help reduce corrosive action. By varying the amount of chloride and water solution, freezing temperature can be adjusted from $32°F$ to as low as $-60°F$. Different concentrations of salt are used depending on whether the holding plate is used for a freezer compartment or a refrigerator box. Plates used for refrigeration are designed to freeze at $18°F$ to $26°F$. Freezers use plates which freeze at $-6°F$ to $-25°F$. Table 10.2 shows freezing temperatures for various solutions of water and calcium chloride.

Holdover plates which use calcium chloride brine are constructed from steel tubes and containers. The holdover plates are evacuated of air and sealed. By eliminating the oxygen inside the plate, corrosion is reduced. Chromate additives are also introduced to reduce corrosion. The *ph factor* of the brine is adjusted to be slightly alkaline.

Freezing Temperature Fahrenheit	Percent Glycol	Specific Gravity
20	32	1.05
10	40	1.07
0	43	1.075
-10	45	1.08
-20	50	1.09
-30	53	1.096

Table 10.3: Glycol Brine

While calcium or sodium chloride brines are very efficient, they are corrosive. To evacuate oxygen from them requires thin and strong plates. Other solutions are used which are not as corrosive. These solutions can be used in plates with more internal volume, since the mechanical stress is not so severe. Volume is required to provide sufficient time between pump downs. Glycol is a solution which is used by many *do it your selfers* in place of calcium chloride.

Glycol is better known as anti-freeze. Glycerin, ethylene glycol or propylene glycol are used. Glycol is heavier than water but not as heavy as calcium chloride. Glycol brine therefore does not carry as many *Btu* per pound as does calcium chloride. In fact, about 35% more volume is required for the same *Btu* storage. Table 10.3 gives data on ethylene glycol and water.

The use of glycol has other side effects other than reduced *Btu* per gallon. Whereas a gelled calcium chloride plate will maintain a relatively constant temperature as it melts, the glycol plate will allow a steady rise in temperature as the water melts. Glycol and water is not a true eutectic solution which freezes and melts at a specific temperature. This is due to the stratification of the brine solution as it freezes. On the other hand, glycol is readily available and is non-corrosive. Plates using glycol do not need to be evacuated. Aluminium, copper, tin or stainless steel can be used for plate fabrication. For these reasons, glycol and water are the common choice of persons building their own systems. While you can save a few dollars by building your own holdover plate, remember that you must add another 35% in volume, and be willing to tolerate wide temperature swings in the box as the solution is frozen and later when it melts.

A solution of alcohol and water can also be used as a holdover brine. Alcohol is lighter than water so a larger volume is required to get the required weight. The specific heat of alcohol is 0.6, so as more alcohol is added to lower the freezing point, the latent heat value falls as the volume of solution increases.

A 50% solution of alcohol and water will freeze at -20°*F*. The solution at that temperature has a specific gravity of 0.9345. A gallon of it will thus weigh about 7.8 *lbs*. The *Btu*'s of energy stored will be about 900 for the gallon.

Whatever brine is used, it will expand when frozen. If you decide to make your own holdover plate, remember that it cannot be completely filled. The best way to arrive at the right amount of brine is to have a fill cap that is air tight. Initially, fill the plate with more than enough. Now, with the cap removed, freeze the plate. Excess brine will spill from the plate. While the plate is still at its coldest temperature, replace the cap and seal it well. A slight vacuum will be created when the brine thaws. This method is not recommended for plates which are not strong enough to withstand the vacuum.

10.10 Summary

In this chapter, the basics of compressor refrigeration have been presented. Pertinent parameters of compressors, evaporators, condensers and metering devices have been explained. Auxiliary components such as heat exchanger/accumulators, filter/driers, liquid sight glasses, and pressure limit switches have also been discussed. The aspects of continuous cycle and intermittent cycle refrigeration have been presented.

Chapter 11

Non–Compressor Refrigeration

11.1 Thermoelectric Refrigeration

Thomas Seebeck discovered in 1821 that an electrical circuit made of two dissimilar metals would generate current flow if the two metals were maintained at different temperatures. The amount of current that flows is a non-linear function of the temperature difference between the two metals, and in modern times a widely used method of temperature measurement. Two dissimilar metals used to measure temperature are called thermocouples.

Seebeck did extensive classification of materials based on the current flow and temperature differences but failed to grasp the significance of his discovery. In 1834, Peltier observed that current passing through a junction of dissimilar metals would heat one of them and cool the other ... the opposite of the Seebeck effect. Peltier didn't realize the significance of his discovery either. In any case the effects between metals is relatively small and the idea that you might heat or cool from the effect is not readily apparent.

With the advent of the transistor in 1948, greater Seebeck effects became possible. By carefully selecting semiconductor characteristics, devices can be fabricated which exhibit useful heating and cooling. Figure 11.1 shows a thermoelectric module connected as a heater. A module is constructed of semiconductor P type material and N type

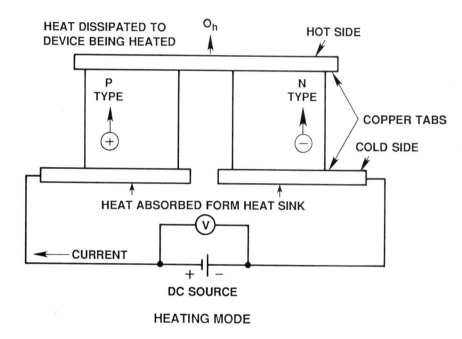

Figure 11.1: Thermo–Electric Heater

material. One end of the P and N material is joined by a metal plate. When current is passed through the module the common plate either absorbs heat or dissipates heat. Whether it absorbs or dissipates depends on the direction of the current flow.

Figure 11.2 shows the same module connected to cool rather than heat.

The amount of heat conducted at the common plate is related to the difference in temperature between the plates and the square of the current flowing times the electrical resistance of the module legs. This allows the amount of cooling to be adjusted by controlling the amount of current flow. This can be done with solid state electronic devices. Thus, refrigeration can be obtained which has nothing moving but electrons! No pumps, compressors, or motors to wear out or break.

A significant advantage of the thermoelectric module is the com-

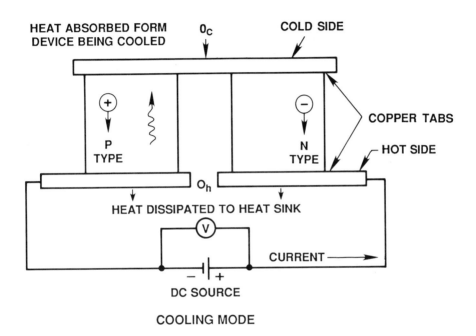

Figure 11.2: Thermo–Electric Cooler

plete lack of moving parts to make noise or wear out. Whereas the compressor condenser and evaporator are heavy and bulky the thermoelectric module is compact and relatively light. And thermoelectric modules are quite inexpensive. So why aren't they used more?

As we noted earlier, a compressor system will typically yield 1–3 times the cooling *Btu*'s that are used to drive the compressor. That is to say that if 100 *Btu* equivalent of electrical energy is necessary to rotate the compressor, the system will yield 100 to 300 *Btu* of cooling

A refrigeration system based on thermoelectric modules is not nearly this efficient. In fact, the numbers are just reversed. To get 100 *Btu* of cooling requires an input equivalent of 150–300 *Btu*'s. With electrical energy as scarce as it is in an alternate energy system, the thermoelectric module has limited, if any, application. We can only hope that thermoelectric module manufacturers find a breakthrough and approach theoretical performance.

(If you must have a very small amount of cooling, such as adequate for storage of medicine, then you might consider a thermoelectric unit. Don't plan to cool too many beers with one unless you have plenty of solar panels.)

11.2   Absorption Refrigeration

In 1824, Michael Faraday was performing experiments in an attempt to liquefy certain gasses which were believed to be *fixed* in a gaseous state. He exposed silver chloride to dry vapors of ammonia until the silver chloride had absorbed large amounts of it. Then he sealed the compound in a test tube shaped as an inverted V. One leg of the test tube was put in cool water, while the other leg was heated. Soon blue liquid ammonia was condensed in the leg which was cooled by water.

After removing the heat from the other leg, Farady observed that the ammonia boiled furiously. As the ammonia evaporated it made the test tube frosty cold.

Absorption refrigeration operates around these principles. First a gaseous refrigerant is absorbed into another compound. Then the compound is heated to drive out the vapors under pressure. The vapors are condensed into a liquid and then evaporated to produce cooling.

An elementary absorption system is shown in **Figure 11.3**. Two tanks are connected by a pipe. One tank has ammonia while the other contains water. Left standing, all the ammonia will evaporate and the water will absorb the vapors. As the ammonia evaporates, it cools its surroundings. Once the water has absorbed all the ammonia possible, cooling stops. The system must be recharged. To recharge the system, heat is applied to the water. Heating generates ammonia vapor. If this vapor is cooled in the other tank, it will condense back to a liquid. When all the liquid is recaptured, the heat can be removed from the generator and the evaporation/absorption cycle repeated.

In the earlier part of the century, domestic refrigerators were built around this idea. Kerosene burners were used to heat the water and generate ammonia vapors. Once a day for 20–40 minutes the system was heated to produce liquid ammonia. For the remainder of the day, the ammonia evaporated, cooling a small box.

Commercial application of absorption principles use lithium bromide as the absorber. Such systems also use a mechanical pump to recirculate the absorbed solution back to the high pressure of the generator. (The generator operates at about 200Ψ.) Continuous operation is possible, using the mechanical pump.

Even the mechanical pump can be avoided by using hydrogen in the system. The AC/DC/Propane units found in RV's operate in a continuous generation/absorption cycle. Hydrogen circulates on the evaporator side but is prevented from reaching the condenser by a liquid filled trap (just like a sink trap). The presence of hydrogen at the same high pressure as the generator side means that the liquid ammonia can flow from the condenser to the evaporator under simple gravity.

The lack of mechanical parts in the hydrogen/ammonia absorption system make it attractive. It can work from any heat source, including waste heat from engines. What are its drawbacks? First, commercial units operate from AC, DC or propane. None of these fuels can be considered abundant in the alternate energy system. To store large quantities of propane requires a large volume. If absorption systems were available that ran on simple diesel fuel they might well become the refrigerator of choice.

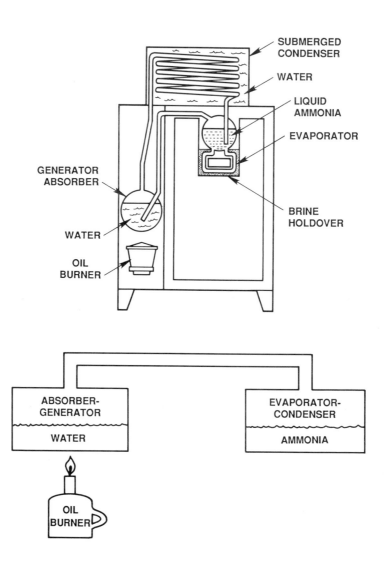

Figure 11.3: Intermittent Absorption Refrigerator

There is at least one other drawback to the absorption system. They must be kept relatively level to operate correctly. If not kept level the trap between the condenser and evaporator may allow hydrogen to pass, backing up into the condenser and preventing proper condensation of ammonia. Absorption type units also employ a flat plate evaporator which is not very efficient unless level. Despite the advances made to alleviate leveling requirements, the absorption refrigerator is not suited to a sailboat that may spend many days on a heel sloshing to windward.

As usual, if we are to operate the absorption refrigerator from our *DC* electrical system, questions regarding overall efficiency must be answered. Unfortunately, the system doesn't quite break even. As in the case of thermoelectric modules, more energy has to be put into the system than is obtained in cooling effect. Input energy is 1.25–2 times that of cooling. When the generator is heated from propane, additional inefficiencies are incurred. Much energy is wasted up the chimney. Fired directly by fuel, 2–4 times as much energy must be put into the generator as is obtained from the evaporator.

If propane is readily obtained and inexpensive then absorption refrigeration may be right for you. If you must rely on *DC* power to drive the refrigerator then stick with the compressor.

11.3 Summary

In this chapter thermo-electric refrigeration and absorption refrigeration have been presented. The efficiency of neither method is such to recommend their usage in the alternate energy system except for very small scale applications.

Chapter 12

Refrigeration Calculations

12.1 General Information

In the prior material, we have covered refrigeration in general and attempted to provide an intuitive grasp of the essential facts that govern this technology. That knowledge alone will suffice for many, allowing more efficient refrigerator use.

This chapter is for the true masochist who wishes to design his or her own refrigeration system. In this chapter, the gas tables for R–12 are presented, and their use explained. In addition the calculations to determine refrigeration load are given, along with the calculations to properly size a compressor and condenser to the load. We also delve into such things as the rate that a condenser or evaporator can conduct heat, the importance of refrigerant flow velocity, and, *how fast a holdover plate can be pumped down.* Even if you don't want to design your own system, you may benefit from this chapter. The mathematics is limited to simple algebra, but even if you don't work through the examples, there is much useful information to be gained.

As we know now, the refrigeration system is principally comprised of the compressor, the condenser and the evaporator. The liquid metering device and tubing are other parts requiring attention to detail. For each of these a coefficient of performance must be obtained before overall system operating parameters can be known.

The design of a refrigeration system begins by doing a load analysis. The load analysis will describe the amount of cooling which must be done by the evaporator. This in turn will dictate the amount of refrigerant which must be evaporated. For a given quantity of evaporation, the load on the compressor and condenser can be calculated. By knowing the load on the compressor, the energy to drive it can be calculated. Likewise, the load on the condenser can be translated to energy which must be expended to keep the condenser at operating temperature. Such is the design path for the refrigeration system, and one that must be accomplished prior to the design of any electrical system which supports refrigeration.

A design can begin by treating the major symptoms. By that we mean plugging the biggest holes from which energy escapes. This is about the level of engineering that goes into domestic refrigeration. In the alternate energy system, we have to count all the *Btu*'s. Compromise is a word which must not be in our vocabulary.

12.2   The Gas Tables

Given in Table 12.1 is the pressure, volume and heat values for R–12 at various operating temperatures. The information in this table is used to determine refrigeration effect of the evaporating refrigerant, and indirectly the power required to compress the vapors. Examples of using the Table 12.1 follow. *Enthalpy*, seen in the table is a nice word meaning heat content and is measured in *Btu* per pound.

Refrigerating effect stated in *Btu* per pound of circulated refrigerant is given by subtracting column 5 from column 6 at the two temperatures of the condenser and the evaporator. For instance, assume an evaporator temperature of -10°F. The refrigerant *vapor* enthalpy at that temperature is 76.2 *Btu* per pound. Assume a 100°F condenser. The *liquid* contains 31.1 *Btu* per pound at that temperature. The net refrigerating effect is 76.2 - 31.1 = 45.1 *Btu* per pound. To get 451 *Btu* per hour of refrigerating would require that 10 pounds of refrigerant per hour circulate through the system.

At -10°F, the vapors occupy 1.97 feet^3 per pound, so 10 pounds will occupy 19.7 feet^3.

Consider now an evaporator at 20°F, and the same 100°F condenser. Subtracting 31.1 from 79.4 leaves 48.3 *Btu* per pound. To

Temp	Pressure	Volume	Density	Enthalpy	
(F)	(Ψ)	(cu ft/lb)	(lb/cu ft)	(Btu/lb)	
	Absolute	Vapor	Liquid	Liquid	Vapor
-40	9.3	3.88	94.7	0	72.9
-35	10.6	3.44	94.2	1.1	73.5
-30	12.0	3.06	93.7	2.1	74.0
-25	13.6	2.73	93.2	3.2	74.6
-20	15.3	2.44	92.7	4.2	75.1
-15	17.1	2.19	92.2	5.3	75.7
-10	19.2	1.97	91.7	6.4	76.2
-5	21.4	1.78	91.2	7.4	76.7
0	23.9	1.61	90.7	8.5	77.3
5	26.5	1.46	90.1	9.6	77.8
10	29.3	1.32	89.6	10.7	78.3
15	32.4	1.21	89.1	11.8	78.9
20	35.7	1.10	88.5	12.9	79.4
25	39.3	1.00	88.0	14.0	79.9
30	43.2	0.92	87.4	15.1	80.4
35	47.3	0.84	86.9	16.2	80.9
40	51.7	0.77	86.3	17.3	81.4
85	106.5	0.38	80.8	27.5	85.7
90	114.5	0.36	80.1	28.7	86.2
95	123.0	0.33	79.5	29.9	86.6
100	131.9	0.31	78.8	31.1	87.0
105	141.3	0.29	78.1	32.3	87.4
110	151.1	0.27	77.4	33.5	87.8
115	161.5	0.25	76.7	34.8	88.2
120	172.4	0.23	75.9	36.0	88.6
125	183.8	0.22	75.2	37.3	89.0
130	195.7	0.20	74.2	38.6	89.3
135	208.2	0.19	73.6	39.9	89.7
Col 1	Col 2	Col 3	Col 4	Col 5	Col 6

Table 12.1: Saturated R–12 Properties versus Temperature

achieve the same 451 *Btu* per hour means that only 9.34 pounds of refrigerant need be circulated. The volume of vapor at $20°F$ is 1.1 foot^3 per pound. Therefore, 9.34 pounds will occupy about 10.3 feet^3. Note that the compressor must compress 19.7 feet^3 at $-10°F$ to achieve 451 *Btu* per pound, but only about half as much (10.3) at $20°F$. This exercise should convince you to keep your box as warm as practical, since lower temperatures require a significantly larger volume to be compressed. The same compressor will have to run about twice as long to achieve $-10°F$ rather than $20°F$. Usually, compressors operating with low temperatures are rotated faster to achieve the extra volume necessary. This is not without its own set of problems.

Observe the rapid rise in pressure that occurs whenever condenser temperatures get above $100°F$. If we allowed the condenser to rise to $130°F$, and tried to maintain a $-20°F$ evaporator, then the compressor would have to work with a compression ratio of 12.8 to 1. (195.7/15.3). The practical limit for a compressor is about 10:1 and at the upper limit, volumetric efficiency falls off dramatically. A single compressor will not be able to maintain $-20°F$ with a condenser temperature of $130°F$.

Volumetric efficiency is the ratio of actual compressor performance relative to the theoretical volume per piston stroke. Actual performance is always less than theoretical. Some vapors are lost past the piston. The piston cannot travel all the way to the head, so some volume is lost to clearance space. Valves do not open and close instantaneously and that incurs additional losses. At higher compression ratios even the kinetics of pulsating vapors decreases performance. Figure 12.1 shows volumetric efficiency versus compression ratio. Note that at a 10:1 ratio, only about 50% of theoretical performance is achieved. This figure will be even lower for a compressor operated at high *RPM*. High compression ratios are the result of very cold evaporators and very hot condensers.

12.3   Load Calculations

Refrigeration load is comprised of two major components. One is the useful load, that which we wish to make cold, while the other com-

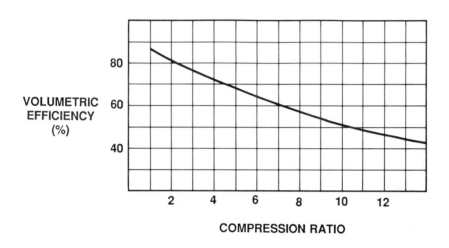

Figure 12.1: Volumetric Efficiency versus Compression Ratio

Item	Btu
Insulation Loss	143
Door Open/Closing	66
Water	35
Other Refreshments	35
Foods	35
Total	314

Table 12.2: Refrigeration Load, Btu per Hour

ponent is the overhead load. Overhead load is made up of insulation losses and other losses such as air leaks or other conduction through water drains, etc. Losses due to insulation have been previously described, and the arithmetic to work the problem was given earlier. As noted, the biggest refrigeration loss is the insulation.

In a well sealed box, the major loss due to air currents will be the actual opening and closing of the refrigerator. A top loading box will have little energy lost compared to the front opening boxes typical of domestic units. Consider a 12 foot^3 refrigerator that is 50% full of food. That leaves 6 feet^3 of air which will drop out of the box every time the door is opened. If that air is relatively dry, 6 feet^3 will weigh about half a pound. The specific heat of dry air is about 0.24. Every time the door is opened, you will lose 16.5 Btu. If the air is moist, much more energy will be lost. How many times a day will the door be opened? Four times an hour is 66 Btu's, nearly half a pound of ice. Do you still want a box that opens from the front?

The useful load will depend on the amount of foodstuff which is transferred through the box regularly. In hot weather, a gallon of fluid per day, per person is not unusual. With two people drinking out of the box, that is two gallons a day required. This problem was worked earlier, and under the conditions, amounted to 42 Btu an hour. If the liquid cooled is in bottles, then refrigeration is also required for the glass itself. Specific heat of glass is 0.18. Cooling 5 pounds of glass from $100°F$ to $40°F$ will consume 54 Btu. $(5)(0.18)(60) = 54$.

Other foods must be cooled. Vegetables are from 80 to 90% water. The number of *Btu*'s to cool them can be calculated accordingly. Fresh beef and pork is 85–92% water. Fresh fish is 90–95% water. Several pounds per day of these items may go through the box. Eggs are 85–90% water. A dozen large eggs weighs about 1.5 pounds.

In short, you must estimate the daily cooling needs of the useful load and calculate the *Btu* required. To compute refrigeration load add up the individual items. An example is given in Table 12.2.

(Editors Note: Fresh eggs do not need to be refrigerated. Once refrigerated, however, they must remain so. Obtained right from the chicken (before they've been washed) and lightly greased with vegetable shortening, eggs will keep for at least a month if turned over daily to distribute internal moisture. Two Pacific crossings have borne this out.)

In the example, 314 *Btu* per hour is the load on the evaporator. That is the amount of heat that must be transferred from the box through the evaporator. The compressor and condenser must be designed to work into this load plus the inefficiencies of the rest of the system.

If you are designing a holdover system, you can calculate the amount of brine required for the heat transfer of 314 *Btu*. Suppose that you only want to pump down once a day. The holdover plate must store 7536 *Btu* of energy ((24)(314)). If we make a reasonable assumption of 800 *Btu* per gallon of brine for a glycol system, then about 9.5 gallons is necessary. This is a large amount, much more than that provided by the usual holdover plate. But facts are facts ... either we reduce the refrigeration load, pump down more frequently, or provide the brine required. Two or three pump downs per day are not uncommon on many holdover systems in hot weather. If most of your cruising is done in cooler climes, then designing a refrigeration system to meet hot weather conditions may be unnecessary. On the other hand, if you expect good performance in southern latitudes, plan well before you build the system.

12.4   Evaporator Calculations

The above load calculation shows that about 314 *Btu* per hour is necessary. What does this mean in terms of an evaporator for a continuous system such as a *DC* compressor system? In particular, what is the surface area required to conduct 314 *Btu* an hour? In still air, a flat plate evaporator will conduct 2.25 to 2.5 *Btu* per foot^2 per degree per hour.

If we assume a flat plate aluminum evaporator, a box temperature of 40°*F*, and a refrigerant temperature of 25°*F* then the condenser must have a surface area of 9 feet^2 to conduct 314 *Btu* per hour. (314/15 degrees/2.35 = 8.9) Does your evaporator have this surface area? Measure all surfaces of the evaporator. If the evaporator is too small, then the condenser is probably also too small. Maybe even a bigger compressor and motor will be necessary.

12.5   Holdover Plate Calculations

Generally, a refrigeration system is installed after the builder has decided how much space is to be allocated to the reefer. More than likely the box will be too small to start with, and, with inadequate insulation. The buyer, having lost the refrigeration battle already, will be unwilling to give up much additional space for holding plates. Many owners simply elect to use a continuous cycle refrigeration system because there isn't sufficient room in the box for holding plates. That decision means that the modest electrical system installed by the builder is now helplessly underrated.

Suppose for an instance that the reefer was designed as an integral part of the boat, and to reasonable live aboard standards. How would holdover plates then be chosen?

There are three aspects of a holdover plate that are subject to design rules. In particular, we are interested in the *Btu* storage capacity, the cooling capacity and the length of time that it takes to pump down the holdover plate.

As we noted earlier, the length of time that a holdover plate can cool between pumpdowns is directly related to the internal volume and the eutectic solution inside. We assume the hourly refrigeration load has already been calculated ... say 314 *Btu*/hour. How many hours do we want to go between pumpdowns? It is not unreasonable to expect a single pumpdown per day. At least 7536 *Btu* must be stored.

Given the rather poor performance of anti-freeze solutions relative to calcium or sodium chloride, and considering the overall expense of the refrigeration system versus the small amount that might be saved by using a glycol plate, we recommend the use of a chloride plate. Such a plate will store about 1000–1100 *Btu* per gallon, as shown in Table 10.2. It is a simple matter then to determine that we need 7.2 gallons of solution.

The refrigerant tubes and supporting framework inside the plate will consume about 30% of the available volume. This must be accounted for before the actual size of the plate can be calculated. Rounding 7.2 gallons to 10 gallons is appropriate. That translates to 2310 inches^3 which can then be divided up in any suitable dimensions.

It becomes apparent that 10 gallons of solution is a lot when you consider 2 five gallon water jugs. The obvious solution is to provide more insulation and cut the daily refrigeration load. Where that is not feasible, then find the room for 10 gallons.

Besides the storage capacity, you will want to calculate the cooling capacity of the plate. This involves the thermal transfer from the plate to the air, and like the evaporator above, figure on 2.25 to 2.5 *Btu* per foot^2 per degree per hour. If you have an abnormal amount of surface area for the reefer, because you have installed a lot of holdover storage capacity, then the box may get too cold, freezing its contents. Be ready to insulate part of the holdover plate if the cooling capacity is larger than necessary. If you have designed for adequate brine storage capacity, the surface area of the holdover plate will always provide more cooling capacity than is necessary.

Because holdover plates are usually undersized, some manufacturers have gone to a eutectic solution which freezes at 18°*F*. For custom plates which have sufficient storage capacity for 24 hours and a correspondingly large surface area, then a eutectic solution of 24 to 26°*F* is more appropriate. Some local freezing may still occur, but vegetables can be placed high and away from the plate to avoid freezing. Again the plate can be insulated to slow down the conduction of heat from the box.

Freezers pose a special problem. To achieve a fast pumpdown there must be a large temperature difference between the evaporating refrigerant and the eutectic point of the solution. Freezers should be maintained at $0°F$ to $10°F$ degrees to keep ice cream for short periods. This will require a eutectic temperature of $-10°F$ to $-6°F$. The automotive air conditioning compressor is not designed to operate with suction pressures less than atmospheric ... $-21°F$ for R–12. For rapid pumpdown, a temperature of $-30°F$ to $-20°F$ is necessary. While the automotive compressor will not perform well under these conditons, and may in fact void the warranty, some open type compressors can operate in this range.

How fast can you pumpdown a holdover plate? As we indicated earlier, not as fast as would be desirable. In fact, the holdover plate limits the rate at which the system can be pumped down. Whereas a condenser can transfer several hundred Btu per foot^2 per degree per hour, the holding plate is limited to the rate of heat transfer through the tubing inside the plate. The heat conducted is about 18 Btu per foot^2 per degree per hour. The square feet referenced are those of the external surface area of the tubing.

The reason for the low thermal conductivity is the fact that ice forms around the internal tubes and supporting metal work, and ice has a low rate of thermal conduction to the unfrozen brine.

The obvious solution, increasing the amount of tubing to acquire a larger external tube surface area, runs counter to the need for a large volume of brine to provide the holdover capacity desired. The only real way out of the dilemma is to provide much more holdover plate than strictly needed, pumping the plates down in parallel, and frequently enough so that you can freeze the unthawed portion in the time desired. This approach has the side benefit that you may be able to go several days between pumpdowns ... convenient for short vacations. Of course the plates will require a long pumpdown when they become completely thawed.

The only real solution to the slow pumpdown rate of the plate is to give up the idea of a fast pumpdown, and use a compressor which is more closely matched to the rate at which the plate can be frozen. Overall this approach will be more energy efficient. The goal in this case is to make the time spent freezing the plates as painless as possible. Leaving to catch a few bass for dinner is one option that comes to mind.

Another consideration where you have a separate refrigerator and freezer is to use two separate compressors. The two systems operate in parallel so that freeze-up time runs concurrently. With the installation of crossover valves, the systems offer redundant reliability. Compressors, it turns out, are often less than the cost of holdover plates, even including a DC motor to drive the compressor.

Most systems are designed with one or two plates connected in series. As noted above, additional surface area can be achieved by running the plates in parallel. There is a limit to how many plates can be operated in series or parallel. In particular, the velocity of the refrigerant through the plates is important. Plates are prone to trap oil ... to avoid this, high refrigerant velocity is necessary. This means that tubes should not be too large. By the same argument, if the plate tubing is too short, then not only is there a loss of needed surface, but there is a good chance of flooding liquid back to the compressor. These arguments appear to dictate a long, small diameter tube. The length tends to reduce the possibility of flooding the compressor, and yields the desired surface area. The small diameter tubing will maintain a high refrigerant velocity, keeping oil logging to a minimum. As a counter point, however, it isn't desirable to have too much of a pressure drop from input to output, since this lowers the effective plate capacity. As will be seen later, an externally equalized expansion valve is recommended for holdover plates, particularly where input to output pressure is high.

In general, a plate should be operated such that the refrigerant evaporating temperature is 15 degrees less than the eutectic temperature. To prevent oil logging in vertical lines, the velocity of exiting refrigerant should be close to 1500 feet per minute, but not exceeding 1800 feet per minute. A velocity of 700 feet per minute is a bare minimum for horizontal lines at the minimum expected capacity. It is a common misconception that larger tubing will lower the pressure drop from input to output, and therefore increase the capacity. In fact, smaller tubing will increase the pressure and the velocity, but the additional velocity raises capacity more than the increased pressure reduces it. Lines should only be oversized leading to the plate and returning from it. If at all possible, the suction line from the plate to the compressor should make a continuous slope so that no oil is trapped. If that isn't possible, then the vapor velocity must be maintained to carry the oil along with it.

Note again that the temperature difference between the evaporating refrigerant and the freezing temperature of the brine should be about 15°F. It is common practice, when seeking a fast pumpdown, to ignore this caveat. When a large compressor is used, such as an automotive air conditioning unit, the compressor has much greater capacity than the plate can absorb. By allowing a large difference in refrigerant and eutectic temperature, the rate of pumpdown is increased marginally, but in fact the compressor may reach its lower pressure limit before the brine is completely frozen. This situation is analogous to using a giant alternator on a small battery, and expecting the battery to be charged just because the voltage has risen to 14.4 Volts.

All in all, there are many conflicting requirements for the size of tubing in the holdover plate. Finding an optimum balance is not an easy task. Velocity and volume are related by the expression:

$$CSA = Volume/Velocity$$

CSA stands for cross-sectional area of the inside of the tube. The volume is the volume of vapors produced from the amount of circulating refrigerant, so CSA cannot be determined until after the size of the compressor is known.

Assume that we have a compressor which can circulate 2.5 feet^3 of vapors per minute. That becomes our volume.

$$CSA = 2.5/1500 = .00167 \text{ feet}^2 = .240 \text{ inches}^2$$

The diameter[1] of the tube is ...

---

[1]The symbol Π (Pi) is a constant which relates the diameter of a circle to its circumference. The value of Π is approximately 3.14.

$$D = \sqrt{\frac{(CSA)(4)}{\Pi}} = 0.552 \text{ inches}$$

This is the inside diameter of a 5/8 tube. It would be an appropriate size tube for the tubing between the compressor, condenser and holdover plate, and in many cases, the right size tube for the plate itself.

Suppose that parallel plates were desired. Two 1/2 inch tubes will yield a CSA of 0.292 inches^2. With the same vapor volume of 2.5 ft^3 per minute, the velocity will slow to 1230 feet per minute. Three 3/8 tubes will yield a CSA of 0.228 inches^2 and a velocity of 1579 feet per minute.

It should be noted that if you choose to operate series/parallel plate combinations, that one expansion valve is better than multiple valves provided that the total path length of the parallel sections are equal. Don't try to run a two plate series connection in parallel with a three plate series connection.

The total length of tubing inside a plate is also important. While we need to keep the velocity of refrigerant high, the higher the velocity, the higher the pressure drop per foot of length. If we use long lengths of tubing, then the size should be increased to keep pressure drops within limits, even if some velocity is lost. It should be pointed out that friction in a line varies as the square of velocity ... excessive velocity is to be avoided since a large increase in pressure results. The recommended upper limit on a 5/8 line is 32 feet.

Assume that the plate has 30 feet of 5/8 tube. The circumference of that line is $C = (\Pi)(D) = 1.964$ inches. Thirty feet will have 4.9 feet^2 of external surface area. If we limit the temperature difference between refrigerant and eutectic to 15°F, and use 18 Btu per foot^2 per degree per hour, then we can pumpdown at the rate of 1325 Btu per hour. That is probably pretty close to the practical limit, although we might cheat a little and use 20°F as the allowable temperature difference. This would give us a calculated 1767 Btu per hour, but in practice, the k factor of 18 Btu would be less. Perhaps we could actually achieve a pumpdown rate of 1500 Btu per hour by driving the plate colder.

Suppose that we use three plates in parallel, each with 3/8 tube rather than one plate with a 5/8 tube. Repeating the above calculations for one plate and multiplying by 3 yields a pumpdown rate of 2386 *Btu* per hour. This was derived assuming 30 foot lines, 15°*F*, and 18 *Btu*. This is a significant improvement over 1500 *Btu* per hour and clearly demonstrates that if possible, use parallel plates to achieve a faster pumpdown.

In an earlier example, we had calculated refrigeration load to be 314 *Btu* per hour or 7536 per day. If we can only achieve 2386 *Btu* per hour pumpdown rate, then over 3 hours per day will be necessary. If that isn't acceptable, then more parallel plates need be added, with a compressor large enough to drive them. It should be remembered that 314 *Btu* per hour is a lot of refrigeration. Small, well insulated reefers will consume much less than this.

12.6   Condenser Capacity Calculations

As noted earlier, the condenser must transfer from 1.2 to 1.5 times as much heat as the evaporator. This means that in our prior example of 314 *Btu* evaporation, 471 *Btu* of heat must be conducted from the condenser to get 314 usable *Btu* at the evaporator. Heat conduction for the air cooled condenser is calculated using the same 2.25 to 2.5 *Btu* per foot^2 per degree per hour that was used for the evaporator. This assumes still air. Heat transfer can be doubled with 200 feet per minute of air past the condenser, and tripled for 500 feet per minute.

An air cooled condenser transfers heat in proportion to the amount of air flow past the condenser. In still air, 2.5 *Btu* per foot^2 per degree, per hour can be transferred. If we allow a condenser rise of 10 degrees over ambient, then 25 *Btu* per foot^2 per hour means that a condenser of 19 feet^2 is necessary to transfer 471 *Btu*. By moving 200 feet per minute of air past the condenser, the surface area of the condenser can be cut in half. Increasing air velocity beyond 200 feet per minute doesn't decrease the surface area proportionally, as noted above. A six inch fan moving 100 feet^3 per minute of air is 200 feet per minute of air flow. That size of fan is quite noisy and may be objectionable.

A water cooled condenser may have a temperature difference between input water and output water. To derive the temperature difference between the refrigerant inside the condenser, and the outside surface of the condenser, you need to take the mean temperature difference between water input and water output. Remember that dirty surfaces, whether covered with barnacles or dust, reduce heat transfer efficiency. A larger condenser is required for dirty conditions. As usual, there is a tradeoff to be made between the velocity of cooling water, and the surface area of the condenser. The higher the velocity, the less the surface area needed. Higher velocity means more water pumped, however.

Calculating the size of a water cooled condenser is more complex than for the air cooled condenser, but an exercise you must do if you plan to make your own condenser from two tubes. If you buy a condenser you can save a lot of mental exercise and obtain additional operating characteristics from the manufacturer. Water cooled condensers are rated in *Btu* per hour based on 1, 2, or 3 gallons of water per minute. This velocity, incidentally, is about what most of the small engines used in boats produce with their water pumps. Curves on commercial condensers are available that show the entering water temperature versus the condenser temperature. (While we are on the subject, you should note that water always flows through a condenser in the opposite direction than the refrigerant, for maximum heat transfer.)

In a double tube condenser, the inner tube conducts the water. The refrigerant flows through the outer tube, and is thus cooled via the water tube, and by conduction through the outer tube to the atmosphere. Outer tube to air conduction can be calculated on the basis of 2.25 to 2.5 *Btu* per foot^2 per degree per hour as in the case of the air cooled condenser. Generally, this effect will be considerably less than the water cooling, and since it works in favor of a conservative design, it can be ignored.

Water cooling is dependent on the surface area of the outside of the water tube, as well as the velocity of water through the tube. At 50 feet per minute of flow, 185 *Btu* per foot^2 per degree per hour is removed. The temperature difference is that of the water and that of the refrigerant. At 200 feet per minute, 330 *Btu* per foot^2 per degree per hour can be removed. Velocities greater than this do not achieve a proportional increase in heat removal.

On the one hand, we wish to maximize the square feet of surface area that the refrigerant is exposed to. At the same time we wish to have water velocity up around the 200 feet per minute range. Unfortunately, we can't have both unless we are willing to pump an enormous amount of water. Where is the balance point?

Suppose that we need to transfer about 5000 *Btu* per hour through the condenser. Suppose also that we have elected to use a water pump with 3 gallon per minute capacity, and desire a water velocity in excess of 200 feet per minute. To determine the diameter of the water tube we use the formula, CSA = Volume/Velocity. CSA stands for cross-sectional area of the inside of the tube and is measured in square inches. Volume is measured in cubic inches, and Velocity is measured in inches per minute. Converting 3 gallons to cubic inches is a look up exercise using the appendix ... (3)(231) = 693 inches^3. A velocity of 200 feet per per minute is 2400 inches per minute. CSA = 693/2400 = 0.289 inches^2. From CSA we calculate the inside diameter of the water tube.

$$D = \sqrt{\frac{(4)(CSA)}{\Pi}} = 0.606 \text{ inches}$$

This inside diameter falls midway between a 5/8 and a 3/4 inch tube. If we go to the larger size, then the velocity will be less than 200 feet per minute unless we pump more than 3 gallons per minute through it. If we can pump more water, the condenser will be shorter due to the increased surface area of the 3/4 tube. On the other hand, if we use the 5/8 tube the velocity will be higher, resulting in more back pressure on the pump. With the higher velocity, though, we will get more than 330 *Btu* per hour of heat transfer.

Let's work the equations backwards to determine the velocity using the 5/8 tube.

$$CSA = \frac{(\Pi)(D^2)}{4} = \frac{(3.14)(0.555^2)}{4} = 0.242 \text{ inches}^2$$

Using the formula Velocity = Volume/CSA and substituting actual numbers yields ... 693/0.242 = 2865 inches per minute. That is equivalent to 238.7 feet per minute. This is not an unreasonable velocity, so let's choose the 5/8 tube.

The outer diameter is 5/8 or 0.625. The circumference of that diameter is given by $(\Pi)(D) = 1.964$ inches. We started with the

premise that we needed to transfer 5000 *Btu* per hour, and with the
200 feet per minute velocity could transfer 330 *Btu* per foot^2 per
degree per hour. Let's assume that the water temperature is that of
a hot lagoon in Baja California, hovering right at 90°*F*. We wish to
keep the condenser at 100°*F*, so we can only allow a 10 degree rise.
That means we can transfer 3300 *Btu* per foot^2 per hour. The square
feet we need is 5000/3300 = 1.52. This is equivalent to 218 inches^2.
Dividing this by the diameter of the 5/8 tube yields 111 inches or
slightly more than 9 feet in length.

We point out that all these calculations were based on a clean
water tube. For saltwater cooling you must use cupronickel which
by nature is anti-fouling. Nevertheless, the average dirty tube will
only conduct about half of what the clean tube will. You might wish
to use 18 feet instead of 9. Twelve to fifteen feet would be a reasonable choice ... the initial added cost will be negligible compared
to the increased operating power required if the compressor exhaust
temperature/pressure is raised due to an inadequate condenser.

It is possible to go overboard on the condenser, however. With
too large a condenser, and cold water cooling it, then compressor
head pressure may be too low for efficient operation. Under these
conditions, the water velocity can be reduced. Note that centrifugal
pumps can safely be throttled, which actually reduces the energy expended on pumping. By using a centrifugal pump the water velocity
can be adjusted to maintain a condenser temperature close to 86°*F*.
Centrifugal pumps are not self priming, however, and so they must
be mounted below the water line. Even if momentarily exposed to
air, such as might happen if your boat heels too far, the pump can
become air locked. In this case, the condenser will overheat, causing
compressor head pressure to rise. If you have installed a high pressure
cut-out switch on the compressor it will operate to save you a major
failure.

12.7   Compressor Capacity Calculations

The refrigeration load was calculated at 314 *Btu* per hour. This was
multiplied by 1.5 to arrive at a condenser load of 471 *Btu* per hour.

How big a compressor do we need to circulate 471 *Btu* of heat per hour?

The first decision to be made answering this question is how much time during each hour will the compressor run. For greatest reliability, run time should be limited to about 30%, or 20 minutes an hour. That means that when the compressor does run, it must move 3 times as much heat. Thus, the compressor is required to circulate at a rate of about 1400 *Btu* per hour, when running.

When we developed the example with 314 *Btu* per hour, the internal temperature of the box was 40°*F*, and the condenser was 100°*F*. To maintain a 40°*F* box, a flat plate evaporator might have to operate at about 30°*F*. Using Table 12.1, it is possible to determine how much energy must be added to the system to circulate 1400 *Btu* per hour. This is done using column six of Table 12.1 at the two temperatures for the condenser and the evaporator. Find that vapor enthalpy at 100°*F* is 87 *Btu* per pound. At 30°*F*, vapor enthalpy is 80.4 *Btu* per pound. The difference, 6.6 *Btu* per pound is the energy that must be supplied by the compressor for each pound of refrigerant circulated. Before calculating compressor horsepower then, we need to know how many pounds of refrigerant must be circulated.

The task is to translate 1400 *Btu* per hour at 30°*F* evaporator and 100°*F* condenser into a quantity of refrigerant that must be circulated. Referring to Table 12.1 find that liquid R–12 at 100°*F* contains 31.1 *Btu* per pound. This is subtracted from the vapor enthalpy at 30°*F* which is 80.4 *Btu* per pound. The net result is 49.3 *Btu* per pound, which is the refrigerating effect that will be produced for the two temperatures. To obtain 1400 *Btu* per hour of cooling will require circulating 28.4 pounds of refrigerant each hour. Horsepower requirements must be calculated at the rate of 28.4 pounds per hour, even though the compressor only runs for 20 minutes each hour, and thus only meets the 471 *Btu* per hour necessary.

Above, it was shown that the system must supply 6.6 *Btu* per pound of refrigerant. By multiplying 6.6 and 28.4, the system must be able to add about 188 *Btu* per hour. By using conversion tables in the appendix, this amounts to about 0.07 horsepower. This is actual energy added to the system to move the heat. Both compressor volumetric efficiency and motor efficiency must be considered.

Horse Power	Volumetric Efficiency (%)
1/12	40
1/6	55
1/4	65
1/2	72
1	75

Table 12.3: Volumetric Efficiency versus Horse Power

Referring to Table 12.1 again, find that there is 0.92 foot^3 per pound of R–12 at 30° F. Thus 28.4 pounds mean that about 26 feet^3 per hour of refrigerant is necessary. From this, the displacement of the compressor can be calculated. Before doing so, however, the volumetric efficiency of the compressor must be accounted for.

In Table 12.3, volumetric efficiency for various size compressors is given. Because of the small size of the compressor, volumetric efficiency will be low ... on the order of 50%. (Horsepower requirements will double to about 0.14 HP. The compressor is not friction free, so an additional 20% should be added to this number. That yields about 0.2 HP.)

A two cylinder compressor with a bore of 1.5 inches and a stroke of 1.375 inches for each cylinder has an internal volume of 0.0028 feet^3. Reducing this by 50% to make up for volumetric efficiency leaves an effective 0.0014 feet^3. The required volume was 26 feet^3, which translates to 18571 compressor revolutions. The compressor would only have to run at 310 RPM to make this many revolutions in an hour. With only 20 minutes out of the hour to run, the compressor must run at 930 RPM, a reasonable speed. A 1/4 to 1/3 HP motor will be appropriate.

By running 15–20 minutes out of the hour, the total run time per day will be about 6–9 hours. A 1/4 HP motor will draw about 18 Amps. A water pump to cool the condenser may add another 3 Amps. By running 6–9 hours, the refrigeration system will require about 126–190 Ah per day. As we noted much earlier, 314 Btu per hour is a lot of cooling, so the Amp-hours to drive it should come as no surprise.

12.8   Expansion Valve Calculations

The expansion valve meters liquid into the evaporator. Figure 12.2, is
a simplified diagram of the various operating pressures for the valve
and evaporator.

The temperature of the output end of the evaporator is sensed
and the valve reacts to feed just enough liquid into the evaporator
so that all the liquid evaporates before leaving the evaporator. To
operate properly, the expansion valve must be sized according to the
capacity of the rest of the system. A sensing bulb is attached to
the evaporator output. Inside the bulb is a liquid/vapor mix, the
pressure of which is directly related to the temperature of the sensing
bulb. The pressure in the bulb is fed via a capillary tube to one
side of the valve. Increasing bulb pressure tends to open the valve.
Balanced against the bulb pressure is a spring pressure which may
be adjusted, and the pressure in the evaporator. When an expansion
valve with no equalizing tube is used, the pressure in the valve is the
same as the pressure of the evaporator at the input end. If there is too
large a pressure drop through the evaporator, then the valve will not
operate correctly. Earlier, it was recommended to keep a high velocity
through the holdover plate. Doing so will increase the pressure across
the plate, compounding any problems with the expansion valve.

To operate where there is a large pressure difference from input
to output, an external equalizing valve is required. The externally
equalized valve has an additional pressure tube which connects the
evaporator output pressure back to the valve. An extra tube must
therefore be installed in the system. In addition, the externally equalized valve will cost a little more. Obviously, it pays not to use the
externally equalized valve if it isn't required. Figure 12.2 shows an
equalization tube.

In Table 12.4 are the recommended maximum pressure drops allowed with a non-equalized valve. For holdover plate operation at the
colder temperatures, very little pressure drop may be tolerated. In
operation, there is no disadvantage to using an equalized valve, even if

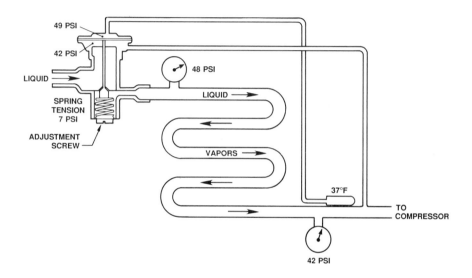

Figure 12.2: Expansion Valve Adjustment Diagram

Evaporating Temperature,°F	-40	-20	0	20	40
Maximum Pressure Drop, Ψ	0.5	0.75	1	1.5	2.0

Table 12.4: Allowable Pressure Drop vs Evaporating Temperature

the evaporator pressure drop is less than the values given above. Under the circumstances, it makes sense to use an equalized valve if you build your own system, even though you may not need it. By designing the valve into the system initially, you can prevent the possibility of rebuilding a portion of the system later.

Valves are rated in tons of refrigerating effect at a known pressure drop across the valve. A ton of ice melting in 24 hours produces 12000 Btu per hour. ((2000 lbs)(144 Btu) / 24 = 12000). As noted above, the amount of refrigeration used by small systems in on the order of several hundred Btu per hour. Even rapid pumpdowns using automotive air conditioning compressors rarely exceed more than 1/2 ton (6000 Btu). To start with then, expansion valves rated 1/4 to 1/2 ton are the ones of interest to us.

To determine the pressure drop across the valve requires knowing the liquid pressure and the evaporator pressure. These values can be derived from Table 12.1 at the temperatures of the condenser and evaporator. In Table 12.5, a 1/4 ton and 1/2 ton valve are specified at four different temperatures over a range of pressure drops.

Assume that the condensing temperature is 100°F, and the evaporating temperature is -20°F. From Table 12.1 locate the pressure at 100°F and subtract from it the pressure at -20°F. The result is 116.6Ψ. In practice, the pressure at the expansion valve may not be the same as the condensing pressure if the liquid line has any restrictions. A dirty filter/drier can contribute significant pressure loss. Too many fittings or bends in the liquid line can reduce the pressure at the valve. Some pressure can be lost if the liquid line makes a vertical rise from the condenser to the expansion valve. All these losses will be small in a well designed system. Lets ignore those possibilities for the moment and use 120Ψ as the drop across the valve.

Depending on manufacturer, a nominal 1/4 ton valve will have a

cap.	Evaporator Temperature							
in	20°F				0°F			
tons	Pressure Drop Across Valve in Ψ.							
	60	80	100	120	60	80	100	120
1/4	0.25	0.29	0.32	0.35	0.25	0.29	0.32	0.35
1/2	0.50	0.58	0.64	0.71	0.42	0.48	0.54	0.59

cap.	Evaporator Temperature							
in	−10°F				−20°F			
tons	Pressure Drop Across Valve in Ψ.							
	80	100	120	140	80	100	120	140
1/4	0.26	0.30	0.32	0.35	0.23	0.26	0.28	0.31
1/2	0.42	0.46	0.50	0.55	0.35	0.39	0.42	0.47

Table 12.5: Expansion Valve Capacity–Tons

capacity of 0.20 to 0.26 tons at -20°F and 120Ψ drop across the valve. That corresponds to 2400 *Btu* to 3120 *Btu*. If these are not large enough, then going to the 1/2 ton size will be required.

Besides selecting a valve to meet the capacity needed, the charge in the sensing bulb must be selected for the temperature ranges expected. Different gasses are used to provide the pressure changes necessary to control the valve. Some manufacturers have a wide range bulb which can be used with both freezer and refrigerator temperatures. From other manufacturers two different bulb charges may be required. This information must be obtained from the valve manufacturers.

For a valve to operate correctly, good thermal conductivity must exist between the valve and the evaporator output line. The bulb must never be heated, so generally, the bulb is clamped to the suction line close to the evaporator, rather than soldered. Be sure to clamp the bulb firmly in order to assure good thermal conductivity. A silicone heat transfer compound can also be used to assure good thermal contact. Be sure to take measures to prevent corrosion around the junction, since this can result in poor heat conduction.

The expansion valve will need to be adjusted once installed, so be sure to use an externally adjusting type valve. To make the valve adjustment the temperature and pressure of the evaporator outlet will need to be measured. Valve adjustment should only be made after the unit has operated for a few minutes and has reached equilibrium. Generally, the valve is adjusted so that the vapor leaving the evaporator is 6 to 10 degrees above its saturation temperature at the indicated pressure. The amount of temperature rise above the saturation temperature is called *superheat*. For example, from Table 12.1, note that the pressure of evaporating R–12 at -20°F is 15.3Ψ. If the pressure measured at the outlet of the evaporator is indeed 15.3Ψ, then the refrigerant inside is evaporating at -20°F. If the temperature at the outlet is -10°F instead of -20°F, and the pressure is actually 15.3Ψ then some amount of heat (superheat) has been added to the vapor above that necessary to evaporate it. Superheat is required to assure that no liquid returns to the compressor, and results from the fact that all the liquid evaporated some length before reaching the evaporator output.

Figure 12.2 shows the various pressures and temperatures. As shown, the valve is loaded with 7Ψ of spring pressure. Pressure at the input to the evaporator is 49Ψ, while the evaporator output is only 42Ψ. Because the valve has an external equalization tube, the 42Ψ from the output is coupled to the valve and aids the spring pressure. Thus it takes 49Ψ from the sensing bulb to counteract the spring and evaporator pressure. From Table 12.1 find that a pressure of 49Ψ corresponds to a temperature of 37°F. This is the temperature of the evaporator output. Yet the evaporator output pressure of 42Ψ corresponds to a temperature of only 28°F. Nine degrees of *superheat* has been added.

Adjusting the valve may require several attempts. Proceed by measuring the pressure at the outlet. Look up the saturation temperature for the measured pressure, and compare that to the outlet temperature. As noted above, the outlet temperature should be 6 to 10 degrees above the temperature indicated by the pressure measurement, using the pressure to enter Table 12.1.

In dealing with pressure measurements, a word of warning is offered. The pressures presented in Table 12.1 are absolute pressures. Gauges usually read in gauge pressure which is 14.7Ψ below the absoute pressure. For gauge pressure, subtract 14.7 from all the pressures provided in Table 12.1. When you find negative pressures, you will now need to convert to *inches of mercury*. One Ψ is equivalent to 2.036 inches of mercury. Negative presures (vacuums) are usually measured in inches of mercury.

Many system provide service valves at the compressor. It should be noted that the compressor suction pressure will not be the same as the pressure at the evaporator. If you build your own system, install a service fitting at the outlet of the evaporator. Better yet, install an electronic pressure transducer permanently, and dispense with pressure gauge hook-up and purge operations. On-line measurement of outlet pressure can also be used to shut down the system when the plate is frozen.

12.9 Measuring System Performance

It would be nice to be able to precisely measure the cooling capacity of a refrigeration system. To measure the actual cooling capacity of an evaporator would require a heat source which could be precisely adjusted to match the evaporator capacity. This is not eminently practical. As a second option, it is desirable to be able to measure all of the heat that is moved in the system. This heat includes the heat added by the motor and compressor, so an estimate of the evaporator capacity would still have to be made.

With an air cooled unit, measurement of heat flow is difficult, but a water cooled unit can readily be measured. Measuring heat flow is as simple as measuring the temperature rise of the cooling water, and measuring the amount of water pumped.

Assume that you are pumping 3 gallons of water per minute, and the water has a $10°F$ rise from input to output. From the conversion tables, find that 3 gallons is equivalent to 0.4 feet^3. Fresh water weighs 62.4 pounds per foot^3, while sea water can weigh as much as 64.3 pounds per feet^3. Three gallons of fresh water weighs 25 pounds. By raising 25 pounds of water $10°F$ then 2500 *Btu* has been moved. Remember, however that not all of this heat came from the evaporator.

If you are using sea water, while it weighs more per unit volume its specific heat is less than 1 *Btu* per pound. Allow another 20% in your calculations. For example 3 gallons of sea water weighs 25.7 pounds. A $10°F$ rise would indicate 2570 *Btu* moved, but because of the lower specific heat, only about 2056 *Btu* is actually moved from the system.

12.10 Summary

In this chapter, the various calculations necessary to design your own system have been presented. The design process starts with a load determination. Next the size of the evaporators is chosen, followed by a determination of a properly sized condenser. Next, a compressor of adequate bore and stroke is selected, and then the horsepower to drive the compressor is calculated. Finally, an expansion valve of the proper size is chosen.

Item	Value
Latent Heat, water	144 Btu per pound of ice melt
Latent Heat, Brines	800–1136 Btu per gallon
Coefficient of Performance	1 to 3 Btu per Ah equivalent
Volumetric Efficiency	40–75% from 1/12 HP to 1 HP
Condenser Capacity	1.2 to 1.5 Evaporator Load
Condenser Temperature	86°F to 115°F
Condenser Rise	5°F to 15°F
Condenser Pressure	92Ψ to 136Ψ
Evaporator Temperature	-20°F to 40°F
Evaporator Fullness	20% to 30% of liquid.
Conduct; air, still	2.25–2.5 Btu per ft^2 per hour per °F
Conduct; air, 200 fpm	5 Btu per ft^2 per hour per °F
Conduct; air, 500 fpm	7.5 Btu per ft^2 per hour per °F
Conduct; water, 50 fpm	185 Btu per ft^2 per hour per °F
Conduct; water, 200 fpm	330 Btu per ft^2 per hour per °F
Conduct; holdover	18 Btu per ft^2 per hour per °F
Flash Gas, R–12	27% at 86°F liquid to 5°F vapor
k Factor	dry air - 0.16; polyurethane - 0.17
Ice Cream	0°F to 10°F

Table 12.6: Collection of Important Values

Chapter 13

The Balanced Energy System

13.1 Introduction

The point has been made many times that alternate energy is more costly than the power from a public utility. As such, users of alternate energy must first practice wise conservation during power consumption, and secondly, waste as little energy as possible during power generation. By following these practices, the user of alternate energy will find comfort without extravagant costs. Enjoying availability of energy on a full time basis means that one or more forms of energy storage be used. Batteries and refrigeration holdover plates are the two principal storage mechanisms in the alternate energy system. (Land based systems will also utilize energy stored in water for either cooling or heating.)

To design an efficient and economical energy system requires that each part of the energy system be chosen carefully, considering the impact on the system as a whole. No single part of the system can be considered in isolation, because of the interdependence of the parts. The system as a whole must be fine-tuned to provide peak performance, and must be resilient to short overloads.

In this chapter, overall system issues are evaluated for each of the subsystems. We attempt to describe a system which will serve the modest needs of a small mountain home, a medium to large cruising yacht, or *RV*. The design approach presented in this chapter is generally applicable to the design of smaller or larger systems, by scaling the values derived, and following the same design procedure.

No design starts with a requirement, and steadily works toward a fully specified solution. All design is a reiterative process. What seems like a good idea at one stage in a design may be a blind alley that necessitates back tracking. Only after many attempts does a synthesis occur that meets the requirements. While the presentation in this chapter appears to lead from requirement to solution, design in the real world begins as a clutter of wishes and proceeds through a maze of ideas and information. The preceding chapters have presented information which is necessary for a system design. That information will now be drawn upon during the design of the *balanced energy system*.

The first design step is to simply list the requirements or *wishes*. Each item in the list will gain additional specifics as the design proceeds. Some wishes may be tossed aside as impractical. Some wishes may be merged with other features as alternative ways to accomplish the end are discovered. Naturally, as one puts entries into the wish book, it is helpful to know what is possible. In this regard, the prior chapters have shown the level of performance which can be expected from an alternate energy system.

13.2   Setting Requirements

Only you can make a list of requirements for yourself. Below, we take the liberty of describing our own requirements ... perhaps your needs are greater or less. Most certainly they will be different. While we would like to present a cookbook design process that would allow any size or type of system to be designed, such a process is impossible when individual needs are considered. Nevertheless, the process which is used to arrive at our energy system is applicable in general. In reading this chapter, pay more attention to the process than the actual numbers. By understanding the process, you will be able to apply the information of the preceding chapters to your own Balanced Energy System.

In general we want refrigeration and lighting. We want a small freezer and a small refrigerator. From preceding chapters, we know that a freezer will be a substantial load on the system, but nevertheless, we still want one. Additionally, we have a power hungry radar, an even hungrier auto pilot, pumps of various natures, running lights, radios, a personal computer, and even an electric toothbrush. It is only too obvious that our power needs are substantial, and unless we design carefully, our wishes will remain simply that ... wishes.

Besides the basic availability of power and refrigeration, we want the system to be operable over several days with no need for daily charging. This dictates substantial energy storage. On a daily basis, we don't want to run the main engine just to charge batteries or pump down holdover plates. Some alternate means of energy are required, such as a small generator, solar panels or wind generators. The wind and the sun are not predictable, so a small generator in the system is mandatory, but we wish to avoid a daily regimen of generator use. If sun and wind are available, we wish to make maximum use of them.

The fact that we don't want to run the main engine every day means that refrigeration will have to be electrically powered. By choosing electric refrigeration, it should be possible to leave the boat unattended for several days without losing the food in the freezer or reefer. In fact, we *demand* that the system survive by itself for short periods. Cold storage and electrical storage must be adequate to meet this demand, supplemented by automatic generation, when possible.

Our list of requirements is formidable, yet it is representative of the conveniences most of us have come to expect in day to day life. We've camped and sailed in spartan conditions, and admire those who can do so for extended periods. Given our druthers, however, we prefer to live a softer life.

13.3 The Refrigeration System

As indicated previously, the design of the system starts with the loads. Refrigeration is the single biggest load, and must be dealt with before other system components are selected. As noted, both a freezer and reefer are desired.

13.3.1   General Information

The refrigeration system must be able to perform during a Baja summer. Shade temperatures in a breeze can be $100°F$. Surface water temperature can exceed $90°F$ in some protected lagoons, and in the Sea of Cortez itself, water temperature may approach $90°F$. Refrigeration under these conditions, *when it's desired the most* is difficult to achieve unless the system was designed to meet these criteria.

Designing a system to perform well under high ambient temperatures is not difficult, but the components to build it will obviously cost more than the components that go into the usual refrigeration system. Whereas air cooled condensers will work satisfactorily in nothern climates, only a water cooled condenser can be expected to perform well in hot weather. By the same token, a poorly insulated reefer may keep northern beer cool, but fail miserably in a southern anchorage.

13.3.2   Holdover Plates and Freezer/Reefer Boxes

A stated desire was to have storage for *several days*, during which no refrigeration pumpdown is required. Cold storage is directly related to the volume of eutectic solution in the holdover plates. For several days of storage, a lot of volume will be necessary. On the other hand, some compromise must be made on account of physical space limitations. Finding the balance depends on the amount of space you are willing to commit to the refrigerator and freezer. Most, if not all, production boats are built with far too little insulation for the size of boxes used. If you are rebuilding the refrigeration system, you must either add insulation to the insides of the box, losing food storage capacity, or you must enlarge the space outside of the box for additional insulation.

Naturally, we want to spend as little time as possible, pumping down the holdover plates. Note that the limit to a fast pumpdown is the absorption rate of the holdover plate. Applying a big compressor will not pump down a holdover plate any faster than a compressor sized to the absorption rate of the plate. Recall that absorption rate is dependent on the external surface area of the internal tubing.

We have ruled out making our own plates with glycol because they store less latent energy, and will not hold a constant temperature as they freeze and melt. The alternative to making our own plates is to have them professionally made by Dole Refrigeration Company[1]. Because of the way that Dole builds plates, absorption rate can be calculated quite easily from the external surface area of the plate itself. They state that their plates have an absorption rate of about 16 *Btu* per square foot per degree per hour. That is, for every square foot of external plate surface area, with one degree difference between the evaporating refrigerant temperature and the eutectic point of the brine, 16 *Btu* of energy can be stored every hour. To achieve a fast pumpdown requires a large surface area, a large temperature difference or both. Recall that for best results, the temperature difference should be $15°F$, with $20°F$ being the maximum practical temperature difference between the evaporation temperature and the eutectic point of the brine.

To obtain the necessary surface area, and enough eutectic solution for storage, eight holding plates are used ... four in the freezer and four in the reefer. Each of them have the same dimensions, and are driven in parallel from their respective expansion valves. The plates are 14 inches by 13 inches by 2 inches as shown in Figure 13.1. The volume of the plates is 364 cubic inches. Reducing this by 30% for internal structure, which includes the volume taken by the refrigerant tubing as well as the tube supports, leaves a eutectic volume of 255 cubic inches. That translates to 1.1 gallons in each plate. The eutectic point of the brine is different for the freezer and refrigerator.

Freezer Plate Storage Capacity and Load

The eutectic point chosen for the freezer plates is $-6°F$. From Table 10.2, 1 gallon of brine at $-6°F$ holds about 1050 *Btu*. Total freezer storage is thus 4620 *Btu* ... $((4.4)(1050))$.

Inside the freezer, the plates are arranged as a tight cube of 14 inches as shown in Figure 13.2. The plates are 2 inches thick, and there is 8 inches of insulation around the cube, including top and bottom. Based on 48 square feet of external surface area, an insulation k factor of 0.17, and $100°F$ differential temperature from outside the freezer to inside, 94 *Btu* per hour will be lost to the insulation. With storage

---

[1] Dole Refrigeration Company, Lewisburg, Tennessee.

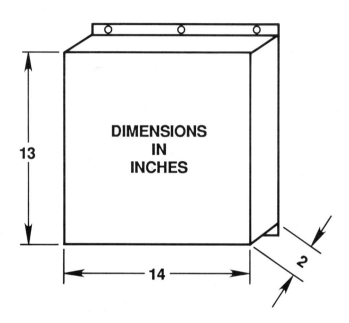

Figure 13.1: Holdover Plate

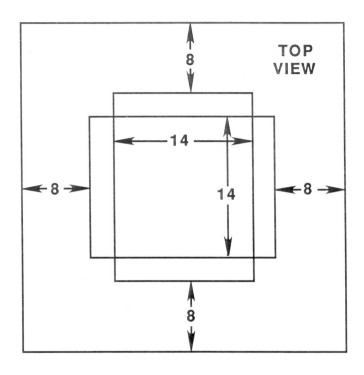

Figure 13.2: Holdover Plates in Freezer

of 4620 *Btu*, the freezer will stay cold for about 2 days assuming no
door seal loss, and no usage load during the period (no door openings
or closings).

In use, the freezer will be called upon to make about 2 pounds of
ice daily, and occasionally freeze several pounds of fish or meat. To
freeze the fish or meat, the compressor will be run extra time, so no
account of this energy is taken as a daily load. Daily usage load to
freeze 2 pounds of ice is 462 *Btu*.

(Two pounds water at $100°F$ cooled to $32°F = 136$ *Btu*. Two
pounds of water freezing $= 288$ *Btu*. Two pounds ice cooled from
$32°F$ to $-6°F = 38$ *Btu*. Recall that ice has about half the specific
heat of water.)

Naturally, the freezer is a top loading box, but there is still some loss due to opening the box. In practice, the box is be opened only a few times daily. At pumpdown time, items to be consumed in the next 24 hours are removed and placed in the reefer where their stored cold is used for cooling. The holding plates inside the reefer are below freezing, so a special place is made for ice and other items from the freezer so that they don't thaw immediately. Assuming that 100 Btu per day is lost opening the freezer is reasonable under normal use.

Note that some frozen foods are placed in the reefer to thaw before they are consumed. This could amount to substantial energy. By the same token, it was assumed above that the water frozen every day came directly from the outside at 100°F. In practice, some or all of the water to be frozen comes from the reefer where it has already been cooled. Because of this swap of energy, and the fact that it is an unknown amount, we have chosen to ignore it altogether in our calculations. The *recovered* energy works in our favor, reducing the actual load on the system.

Freezer usage load is thus made up of the three components of insulation loss, door spillage loss, and the useful load of making two pounds of ice. The total load amounts to 2818 Btu ... ((24)(94) = 2256 + 462 + 100 = 2818). As expected, the freezer is an energy hog. With only 4620 Btu of storage, the freezer must be pumped down at least every 39 hours when operated under normal load at the high temperature of 100°F.

Reefer Plate Storage Capacity and Load

The eutectic point chosen for the reefer plates is 26°F. At that temperature about 1140 Btu per gallon will be stored. The reefer plates store 5000 Btu.

Figure 13.3 shows the way that the four plates are arranged inside the reefer. Four inches of insulation is used. Based on the box dimensions shown, a k factor of 0.17, and a 70°F temperature differential, the insulation loss will be about 50 Btu per hour, or 1200 Btu per day.

In use, the reefer must cool about 2 gallons of fluid on a daily basis. Assuming the fluid is cooled from 100°F to 35°F, then 16.7 pounds of water must be cooled 65°F, using 1085 Btu. That is equivalent to about 45 Btu per hour. Other loads, including losses due to door

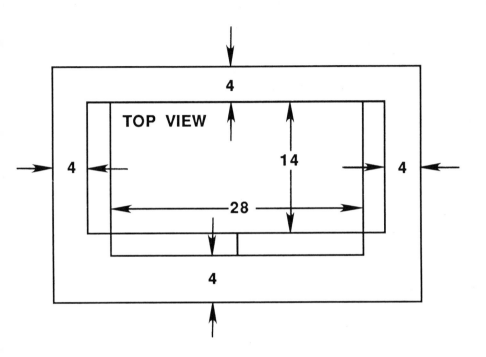

Figure 13.3: Holdover Plates in Reefer

openings will incur another 25 *Btu* per hour, thus, estimated usage load per hour is about 70 *Btu*. Total reefer consumption is 120 *Btu* per hour. With 5000 stored *Btu*, the reefer needs to be pumped down at least every 42 hours. With no activity, the reefer can survive for 100 hours ... a little over 4 days.

Storage and Load Conclusions

From the foregoing, the objective of several days cold storage has been partially met. The freezer lasts 2 days without a pumpdown, and the reefer for 4 days, with no activity. For longer periods of time, either the system must be made to be self tending, or additional cold storage must be obtained. The system can be made self tending by a daily timer that turns the system on at a set time. Turn off is accomplished by sensing the low pressure from the evaporator, the normal way that the refrigeration compressor is shut off. With automatic start, unless batteries are recharged as necessary, there will be dead batteries and a warm freezer/reefer eventually.

Additional cold storage can also be obtained for a planned absence. To extend the time that the reefer can go without a pumpdown is easy ... make as much ice as the reefer will hold. Extending the freezer storage time requires a eutectic solution which freezes/melts around -6°*F*, just like the brine in the holding plates. Brine tanks without the usual plumbing for refrigerant coils are available. Freezing one of them just prior to a vacation could be sufficient to keep the freezer functional without a pumpdown. *Btu* capacity of the brine tank is calculated in the same manner as the holdover plate, except no allowance is necessary for internal structure.

Absorption Rate

Each holding plate is 14 by 13 by 2 inches. External surface area is thus 3.28 square feet. For every degree of temperature difference between the refrigerant evaporating temperature and the eutectic point, 52.4 *Btu* per hour can be absorbed by the plate. Assuming that 15°*F* of temperature difference is maintained, then each plate can absorb 787 *Btu* per hour. With 4 plates, that amounts to an absorption rate of 3147 *Btu* per hour.

The daily freezer load is 2818 *Btu*, while the daily reefer load is 2880 *Btu*. Remember, these loads are under full use with ambient temperature at 100°*F*. On the average, even in Baja California, the actual loads will be less. Then again, there may be more thirsty guests around to help drink the cold water and beer. In any case, the holdover plates can absorb the daily *Btu* requirement in less than one hour each. Because the reefer eutectic point is 26°*F*, an evaporating temperature of 11°*F* will suffice to achieve a 15°*F* difference. The freezer eutectic point is -6°*F* so the refrigerant must evaporate at -21°*F*. To make calculations a little simpler, assume that the freezer evaporates at -25°*F*, and the reefer evaporates at 5°*F*.

13.3.3   Current Design Summary

At this point in the design, the box capacities have been fixed as shown in Figure 13.2 and Figure 13.3. The amount of insulation has been fixed, and the *Btu* losses through the insulation have been calculated. Losses due to door openings have been estimated, and the useful refrigeration load has been specified. Holding plate volume is known, as well as the amount of *Btu* storage provided. In addition, the absorption rate of the holdover plates is known.

The next questions that must be answered regard the issue of compressor capacity. How big a compressor is required? Should there be one or two compressors?

13.3.4   Motors/Compressor Selection

Given in Table 13.1 are approximate compressor capacities relative to the power required to drive them. Also given is the amount of Amps necessary under typical conditions to drive the compressors. Twelve Volt motors are assumed. The temperatures given in parenthesis are those of the evaporating refrigerant. As noted above, the eutectic point of the brine for the freezer is -6°*F*, and for the reefer, 26°*F*.

If the compressor is chosen to match the absorption rate of the holding plates, then a 3/4 *HP* unit would be chosen for the freezer, and a 1/2 *HP* unit for the reefer. With these two compressors, both systems would be operated at once, and the daily refrigeration load

	Freezer($-25°F$)		Reefer($5°F$)	
HP	Btu/hour	Amps	Btu/hour	Amps
1/3	1100-1300	20	2400-2500	24
1/2	1600-1850	30	3300-3600	35
3/4	2700-2900	48	5000-5400	54

Table 13.1: Compressor Capacity

could be met in less than one hour. Note that a single compressor, such as an automotive air conditioning unit could supply both the freezer and reefer plates, also in less than an hour of engine time, if the main engine were used to power the compressor.

The engine has already been ruled out, however, because we don't want to run the engine on a daily basis. Recall that we wished to make the best use of solar and wind power, mandating the use of electrical drive for the compressor(s). To best use either wind or solar energy sources, the source should be coupled directly to the compressors, using the batteries as little as possible for the refrigeration system. Considering the Amp draw of the various compressors, and the potential Amps available from either solar or wind generation, then the 1/3 HP unit makes the most sense, when sun or wind powered.

Added to the Amp draw of the motors listed in Table 13.1, there is some energy required for pumping condenser water. The faster we wish to pumpdown the refrigeration system, the more water we will have to pump through the condenser to keep it cool. At this point, the amount of cooling water has not been determined, but we should plan on 3-5 Amps of current for the water pump when we are estimating total electrical draw.

While the solar panel array and/or wind generator may not be able to produce all of the necessary energy (23-29 Amps), it is desirable to come as close as possible.

The 1/3 HP unit can supply the freezer load of 2818 Btu in about 2-2.5 hours. The reefer load can be acquired in a little over an hour, using the 1/3 HP motor. Therefore, the compressor will have to run about 3.5 hours to pumpdown both the freezer and reefer.

By using the 1/2 *HP* unit, the freezer can be pumped down in about an hour and 40 minutes, while the reefer can be pumped down in 45–50 minutes. Total run time would be about 2.5 hours for the 1/2 *HP* motor.

While it may be possible to get 23 Amps from a solar array, just enough to drive the 1/3 *HP* compressor for the freezer, the 35 Amps necessary to drive the reefer with a 1/2 *HP* unit looks like more than can be expected from the sun or wind.

Another alternative which must be considered is the use of two compressors. When the sun or wind is powering the system, only one compressor at a time is run. If an engine is needed to power the system, then both compressors can be run together to achieve a quicker pumpdown. If two compressors are used, then a good argument can be made for powering the freezer with a 1/2 *HP* motor, and the reefer with a 1/3 *HP* motor. Running the two in parallel will require about 54 Amps, and pumpdown can occur in about an hour and 40 minutes. If the freezer is driven with 1/3 *HP* instead of the 1/2 *HP* motor then pumpdown takes about 2.5 hours.

It can be seen that considerable compromise is required no matter which compressor unit is chosen. While the pumpdown times are reasonable for the 1/2 *HP* unit, the Amps required are more than can reasonably be expected from the sun or wind. That means an engine must be run, or batteries discharged. Even the 1/3 *HP* unit will stretch the sun and wind resources to the limit, and, with only 1/3 *HP*, pumpdown times stretch out to exasperating limits if an engine must be used to power the system. For sun or wind power, the optimum solution would be a small motor/compressor that can be run for 6 to 8 hours during the day. When an engine must be run to provide the energy, then the compressor should be as large as required to meet the absorption rate of the holdover plates.

Without some additional information, a decision about which system is most functional is simply a matter of taste. As will be seen later, however, we have 40 Amps of charging power available from a small generator. This is enough to power the 1/2 *HP* motor, including the water pump for condensing water. Using this generator, the total pumpdown will be about 2.5 hours if both compressors are 1/2 *HP*. By using two compressors simultaneously when the main engine is run, then pumpdown can be accomplished in about an hour and 40 minutes. When only the sun and wind are used, then the batteries will have to make up the difference between the energy source and the immediate refrigeration demand.

The refrigeration system therefore is two systems. A dual system offers redundancy of sorts. If the reefer should break, then it is possible to freeze ice in the freezer and transfer it to the reefer. If the freezer breaks, eat the ice cream first. To achieve true redundancy, the remainder of the system components should be identical as well If constructed in this manner, it is possible to install crossover valves so that either compressor can cool either plate.

Motor *RPM* and Power Calculations

Both compressors chosen for the system are the same model, but driven at different *RPM*. The freezer compressor must be operated faster than the reefer compressor to be able to compress the necessary volume of refrigerant. The bore and stroke on the compressors are 1–7/16 by 1–3/8. For each revolution 0.00258 cubic foot is compressed. The maximum allowable compressor speed is about 1000 *RPM*. At 950 *RPM*, 2.45 cubic foot per minute can be compressed. In an hour, this amounts to 147 cubic feet. Assume that the volumetric efficiency is 70%. That leaves 103 cubic feet actually compressed. Referring to Table 12.1, at -25°*F* evaporating temperature, find that there is 2.73 cubic feet per pound of refrigerant. The compressor can circulate 37.7 pounds of *R-12* per hour. With a 105°*F* condenser and -25°*F* evaporating temperature, the refrigerating effect is 42.3 *Btu* per pound for a total of 1600 *Btu* per hour. On the average, the condenser runs cooler, and the evaporating temperature doesn't get to -25°*F* until the plate is almost frozen, so actual hourly capacity is higher. Volumetric efficiency may also be higher than 70% ... perhaps as high as 75%. As noted in Table 13.1, the expected capacity is 1600 to 1800 at -25°*F*, which is right in line with our calculations.

Repeating the same calculations for an evaporating temperature of 5°*F* for the reefer (19°*F* less than the eutectic temperature) yields a compression capacity of 70.5 pounds per hour. The refrigerating effect of this amounts to 3210 *Btu* per hour. Volumetric efficiency is slightly higher for the higher temperature, however, so more output results. Unfortunately, the holdover plate is not able to absorb it.

To calculate the horsepower required to drive the freezer compressor, find that at -25°F, the *Btu* per pound is 74.6. At 105°F, the *Btu* per pound is 87.4. The difference is 12.8 *Btu* per pound. This is the amount of mechanical energy that must be added by the motor to compress the refrigerant vapors. Converting 12.8 *Btu* per pound times 37.7 pounds means that 483 *Btu* of energy per hour must be added by the motor. Converting that to horsepower yields 0.2 *HP*. Mechanical inefficiency of the compressor must be added to this factor. Assume that it looses 20% to friction and other losses. Motor *HP* is now 0.25. Motor efficiency must also be added. At 70% efficient, the motor size has grown to 0.36 *HP*. Some energy will be lost to the drive belt. All things considered, a 1/2 *HP* motor is necessary to drive the freezer compressor at 950 *RPM*.

Repeating the same calculations for the reefer compressor yields 9.6 *Btu* of energy per pound which must be added by the motor. At 70.5 pounds circulated and 9.6 *Btu*, the motor must contribute 677 *Btu* per hour. This amounts to 0.266 *HP*. Adding compressor and motor efficiency, yields 0.474 *HP*. This is a little too close for comfort on a 1/2 *HP* motor, particularly when volumetric efficiency may consume another 5%. The effects of aging components must also be considered. To be on the safe side, the reefer compressor should be belted such that it operates about 850 *RPM* instead of 950. This means a different pulley size is used on the motor for the reefer.

The motors are 1/2 *HP*, 12–Volt motors. One *HP* is 746 Watts. Even a cheap *DC* motor will be 70% efficient, so at the worst case, a 1/2 *HP* motor will consume 533 Watts. Typically, the 1/2 *HP* motor will be more efficient. At 80% efficiency, only 466 Watts is necessary.

In summary, the compressors are identical, and both driven by 1/2 *HP* motors. The freezer compressor is operated at 950 *RPM*, while the compressor for the reefer is driven at 850 *RPM*. The freezer unit has a low temperature delivery of about 1600 *Btu* per hour, and the reefer unit about twice as much. Evaporator temperatures were -25°F for the freezer with a -6°F eutectic point, and 5°F for the reefer with a 26°F eutectic temperature. The condenser was assumed to be 105°F. In general, performance is equal to or better than the

244 CHAPTER 13. THE BALANCED ENERGY SYSTEM

calculations indicated due to conservative derating. Recall that the conditions were those of a hot day in Baja California. Most of the time, the refrigeration system has excess capacity, and daily run time is substantially less than that calculated. When it is needed the most, however, the system performs up to expectations.

13.3.5 Condensing Unit and Water Pump

While it is relatively simple to build a *tube in a tube* condenser, the result has a few undesirable side effects. First, the condenser is not cleanable, and secondly, the condenser would not be Coast Guard approved. Since we wish to keep our boat in *charterable* condition, this latter point is important, but not as important as the issue of cleanability. From experience we know that in warm tropical waters, marine life grows fast and wild, even on the best of anti-foulant surfaces. Without occasional cleaning, even cupronickel will accumlate some growth, however dead it might be, and as a result lose cooling efficiency.

The condensers, therefore are commercially available units which can be cleaned from either ends by removing end caps. The smallest unit that is produced in the cleanable variety is rated for 1/3 ton of condensing. Recall that a ton of refrigeration is equivalent to 2000 pounds of ice melt in 24 hours, or 12,000 *Btu* per hour. The condensers are rated at 5000 *Btu* per hour. The freezer evaporator consumes about 1600 *Btu* per hour. The compressor adds about 483 *Btu* to the system, so it is safe to assume that the condenser has to transfer about 2100 *Btu* per hour. About 1.5 gallons of water per minute is required to keep the condenser at 105°*F* with the water temperture at 85°*F*.

The reefer compressor adds about 677 *Btu* to the system, which added to the 3450 *Btu* evaporated yields about 4000 *Btu* that the condenser must transfer. This requires about 2.5 gallons of water per minute at 85°*F*. From the foregoing, a centrifugal pump that can produce about 3 gallons per minute will suffice for either condenser. A throttling valve is advised for the freezer, and even the water flow through the condenser for the reefer can be less than 2.5 gallons per minute if the cooling water is less than 85°*F*.

Because centrifugal pumps are not self-priming, the water intake is located well below the waterline such that normal heeling will not expose the intake to air. Water exhaust is also located below the water line so that no energy is wasted by needlessly elevating the waste water. The water system for the condensers is independent of the engine water system.

13.3.6 Filter/Drier

A liquid line filter/drier is used in both the freezer and reefer systems. Both units are rated at 1/2 ton of refrigeration (6000 *Btu* per hour). This size was not strictly necessary, but by adequate oversizing, pressure drop through the filter is at a minimum.

13.3.7 Heat Exchanger/Accumulator

As noted earlier, no high performance system will be without a heat exchanger/accumulator. While these devices are not inexpensive, they can boost operating efficiency, and save a compressor from damage due to oil or refrigerant slugging back to the compressor on startup. Both freezer and reefer systems use the same capacity unit, rated at about 1/2 ton at an evaporating temperature of -20°F.

13.3.8 Expansion Valves

While it is common practice to use unequalized expansion valves in holdover plate systems, the equalized variety was chosen. From prior chapters, it can be seen that as evaporating temperature decreases, then the allowable pressure drop through the plate must be very small for the non-equalized expansion valve to operate properly. When a plate is warm, having melted all of the brine, then the expansion valve will be full open. During this stage of pumpdown, there may be a considerable pressure drop through the plate. This can cause the valve to close down somewhat which means the plate will not be receiving as much refrigerant as it might absorb. Needless to say, this results in wasted energy.

Because there is no operational disadvantage to using an externally equalized expansion valve, and more precise and constant valve settings can be maintained for maximum plate utilization, both systems use an externally equalized expansion valve. A wide range charge in the sensing bulb was selected, and both expansion valves are the same. There is a capacity difference between the freezer and reefer systems, but the same type of valve with the wide range of charge covers both requirements.

13.3.9   Instrumentation and Control

In the preceding chapter, the subject of performance was discussed, and it was shown that measurement of the temperature rise of the cooling water was a practical way to measure performance. While every system should undergo such an analysis, a high quality system will also incorporate other means to measure performance. We believe that you should be able to tell how well the system is performing at a glance. This requires that the instrumentation be incorporated into the system. By adding instrumentation to the system, it will also be easier to adjust. Naturally, a properly instrumented system is more expensive than one with no gauges at all . . . such is the price of performance.

To make expansion valve adjustments, it is necessary to know the pressure and temperature of the refrigerant at the outlet of the holdover plate. It isn't really necessary to build in either pressure or temperature sensors permanently. It is quite simple to attach a thermometer to the outlet of the holdover plate (next to the sensing bulb of the expansion valve) whenever one wishes to adjust the expansion valve. Sensing pressure at that point is not so easy. While there should be a service valve on the suction line of the compressor, the pressure at that point is not the same as the pressure at the holdover plate outlet. The addition of an accumulator/ heat exchanger causes an additional pressure drop, above and beyond that which occurs through the plumbing. To know what the pressure is at the holdover plate outlet, requires a pressure gauge at that point.

When taking the temperature of the outlet, it is important to make a good thermal contact with the outlet. The temperature probe and outlet should be insulated together so that temperature measurements are not erroneous. Only with an accurate temperature measurement can the expansion valve be adjusted to make maximum use of the holdover plate. If the holdover plate is underutilized, then its *Btu* absorption rate will be less. Adjusting the expansion valve too far open will cause liquid refrigerant to return to the compressor. Neither too little or too much can be tolerated.

By the same token, accurate pressure measurements are necessary for fine-tuning the expansion valve.

In operation, it is also convenient to monitor the temperature and pressure at the outlet of the holdover plate. This data foretells the fully pumped down plate. To make expansion valve adjustment easier, and to acquire operational performance data it was decided to permanently mount a pressure transducer and a temperature transducer at the accumulator/exchanger. The transducer output is read with digital instrumentation so that not only can valve adjustments be made periodically if necessary, but actual performance can also be measured in real time.

It is also convenient to know both the temperature at the compressor outlet, and the temperature at the output of the condenser. This latter temperature is useful to determine the effectiveness of the condenser. In particular, if the compressor head temperature and condenser temperature rise too much, then perhaps the condenser needs cleaning, more water flow, or both. It is also nice to know the pressure in the liquid line, since a high pressure is an indication of a fault somewhere in the system. For that reason, a second transducer is installed in the high pressure line.

The system has two shutoff controls ... liquid line high pressure, and suction line low pressure. Hopefully, the high pressure control will never be called upon to shut down the system, though a clogged condenser, frozen expansion valve, or other system blockage can occur and cause an excessively high pressure. The low pressure control is the primary means of shutting the system down when the plate becomes frozen.

Up until the time that the brine in the plate is almost all frozen, the suction pressure and temperature only change slowly. Once the brine has almost frozen, then the suction pressure begins to fall more rapidly. By observing this fall in pressure, the low pressure cutoff switch can be adjusted accurately. While it is necessary to freeze all of the brine, it is inefficient to continue beyond the eutectic point, and the compressor can easily be damaged if the suction pressure drops too low. Compressor seals can start to leak at very low suction temperature. By having a built in temperature transducer and pressure transducer at the holdover plate outlet, a periodic check on the low pressure cutoff switch is easy.

The last, and vital piece of instrumentation is the liquid line sight glass. We chose a glass that contains a sensor for moisture, indicating excess moisture by a color change.

13.4   The Electrical System Loads

The electrical system loads must be specified before the batteries and charge sources can be determined. Listed in Table 13.2 are the loads for our own boat. As shown, the loads are specified for a worst case day swinging on the hook, as well as a worst case day under sail. Time is specified in hours and the daily *Ah* numbers are in Amp–hours.

Some of the loads are not always used, such as the three cabin fans, but in real hot climates at anchor, they tend to be used extensively if little or no breeze is available. In the Sea of Cortez during the summer, there is usually no breeze at night[2].

The radar is seldom run continuously except when used for harbor transit, or in congested seaways. Typically the radar is left on standby with a *picture* being taken every half hour. Under short crew conditions, however, there are times when only the radar and radar alarm are on watch. Usage of the computer and tape deck varies from day to day, so an average is taken.

It can be seen that substantial electrical energy is used. Nevertheless, the loads are real, and it takes real batteries and charge sources to cope with them. The loads are not excessive, considering the comforts that we have come to expect in a modern cruising vessel. While the use of three fans for 16 hours is not necessary, we have seen times

---

[2]Besides an occasional dip in the water, we discovered that hugging the aluminum mast was a good way to cool off.

Load Name	Amp Draw	Anchor Time	Anchor *Ah*	Sail Time	Sail *Ah*
Refrigeration	40	2.5	100	2.5	100
Anchor Light	2	8	16	0	0
Tri-Color	2.5	0	0	9	22.5
Auto Pilot	6	0	0	24	144
Anchor Windlass	80	0.1	8	0	0
Cabin Fans(3)	3	16	48	0	0
15 W Fluorescent	1.25	6	7.5	12	15
30 W Fluorescent	2.5	6	15	12	30
Fresh Water Pump	7	0.1	0.7	0.1	0.7
Saltwater Pump	7	0.2	1.4	0.2	1.4
Wash Down Pump	10	0.2	2	0	0
Propane Valve	1	1	1	1	1
Spreader Lights	8	0	0	0.2	1.6
Tape Deck	1	3	3	3	3
Radar	4	0	0	6	24
Vhf–Receive	0.5	12	6	24	12
Vhf–Transmit	5	0.2	1	0	0
Loran	0.5	0	0	24	12
Computer	8	4	32	1	8
Total Daily Amp hours			241.6		375.2

Table 13.2: Daily Electrical Loads

when 3 fans is not sufficient to keep the air moving. The computer uses quite a bit of power, but it is necessary, since it is the means of our livelihood. While we would like to get by with less energy, and usually do, the Balanced Energy System is designed to perform under fully loaded conditions.

13.5   The Battery System

Sizing batteries for the Balanced Energy System is easily accomplished in principle. To do so, simply decide how many days you wish to go without the need to charge, and multiply that number by the total daily Ah used. The number you arrive at is only half of the necessary capacity, though, since it is not good practice to discharge below 50% of capacity. As noted earlier, exactly two battery banks are recommended, and both banks should be equal in capacity.

Assume that we desire to go 4 days without charging. At anchor we need 966 Ah of capacity in each bank $((4)(241.6))$. At sea the necessary capacity would be 1500 Ah per bank if we are to observe the 50% discharge rule. Obviously, our desire to go 4 days between charges will have to be relaxed ... 3000 Ah of batteries is not realistic on a small boat. As is often the case, we have settled for the capacity that we can carry.

The batteries are arranged as two banks of 440 Ah each. Each bank is comprised of 4 deep cycle sealed batteries which typically have from 110 Ah to 120 Ah each. Sealed batteries were chosen because of their fast charging ability, and the fact that they can be mounted in any position, and never need maintenance. Under usual conditions, the battery banks are operated on alternating days, with the loaded bank also receiving the charge from the solar panels and wind/tow generator. Used in this manner, one bank gets a 24 hour rest which allows the capacity remaining to be determined. Both banks are topped off whenever the main engine is run for propulsion purposes.

With almost 900 Ah of battery power, there is about 450 Amp hours available between recharges, assuming we adhere to the 50% rule. With no charging from alternate energy sources, the system would need to be recharged every other day at anchor. Under sail, the situation is worse, but, when moving, a tow generator supplies charge.

Because the batteries are sealed units which can be 100% discharged without damage, it is possible to get by with full electrical demand for about 3 days between charges, and still have enough reserve to get the engine started. Of course, if this were done regularly, battery life would be limited to 200 to 300 cycles. Under emergency conditions, the refrigeration system and Tri-Color masthead light can be operated for about 6 days. While 900 Ah is not the capacity that we would like to have, it is ample. With the charging system performing to specification, all of the available Ah capacity can be utilized.

Both batteries are instrumented with voltage and current sensors which are connected to a Monitor/Regulator[3]. Current measurements cover the range from 0.1 Amp to 199.9 Amps, either charge or discharge. By measuring current through both batteries it is possible to observe individual charge currents even when both batteries are being charged at once.

13.6   The Main Engine Alternator

The 900 Ah of battery power consists of sealed batteries which can initially accept from 3 to 5 times their capacity ...we could actually use an alternator in the range of 2700 to 4500 Amps. In case you haven't been to your local boat electrical dealer lately, you may be surprised to find that alternators of that size are not generally available. To be more practical, suppose we limit the size of the alternator to the absorption capacity of the batteries. From Chapter 2, it was noted that a conventional battery can absorb no more than about 25% of its Ah rating. From Chapter 3, we learned that some sealed batteries have a natural absorption rate of 50% of their Ah rating, at an applied voltage of 13.8 Volts. That means we should use a 450 Amp alternator to match the battery absorption rate at a relatively low charge voltage.

Our search for a 12 Volt, 450 Amp alternator turned up nothing. An oil cooled 300 Amp unit made for trans-continental busses was

---

[3]Ample Power Company, Seattle, Washington

Alternator Size	Cold Rating($120°F$) Amps	Hot Rating ($200°F$) Amps
Small Frame	160	131
Large Frame	175	160

Table 13.3: Alternator Specifications

available. It is one beautiful piece of equipment, but it would have cost more than the main engine. Possessing redundant capabilities is a natural idea, so instead of a single big alternator, two alternators serve to *quick* charge the batteries. One of them is a small frame unit which is turned by the same belt that circulates the water pump. The other unit is a large frame alternator which is coupled with two belts to an auxiliary drive pulley connected to the front of the main engine. Specifications for the units are given in Table 13.3.

With almost 300 Amps of charging ability, one hour on the main engine is more than sufficient to accommodate the consumption of 241 *Ah* while at anchor. Under sail 375 *Ah* per day is neccesary, so the running time to restore that amount is about an hour and a half. Whenever the main engine is run, the refrigeration system is pumped down. With both motor/compressors operating, about 80 Amps is used to supply refrigeration power. Depending on the state of the holdover plates, and the state of the batteries, actual charge times vary widely. If a long motor trip is anticipated, charging from the auxiliary generator is skipped, waiting instead to use the main engine. In general, the charging strategy is to use all the available sun and wind power, supplemented by the auxiliary or the main engine. Given all the variables, charging is seldom a regimented operation. With proper instrumentation, managing the electrical system becomes a simple enough task, and is easily accommodated in the daily life.

Both alternators are equipped with Automatic 3–Step Regulators which perform the three cycles of bulk charge, absorption charge and float. The two battery banks are individually sensed for both voltage and temperature. Fail Safe (Isolator) Diodes are used between the alternators and the batteries, so voltage sensing at the batteries is required. The batteries are located outside the engine room, stuffed under bunks. While they don't get abnormally warm, charging after a deep discharge does buildup some internal heat. The temperature sensors automatically lower the charge or float voltage as the temperature rises.

In addition to the two 3–Step Regulators, the small alternator can also be regulated manually by a Multi-Source Regulator. To use the manual regulator, both 3–Step Regulators are turned off. The Multi-Source Regulator was installed before we switched to sealed batteries, and it enabled us to charge fast and full, and also to equalize the liquid electrolyte batteries periodically. Now the manual regulator is only used for very long motor trips, or, whenever most of the available engine power is necessary for propulsion. On long motor trips, we extend battery life by floating at a lower voltage than the setting on the 3–Step Regulator. We have a relatively small engine for the size of boat it drives, so when punching into headseas and/or tidal currents, the manual regulator is used to limit alternator current to just enough to keep up with the immediate loads. By inhibiting any battery charging during these times, most of the engine power goes to the prop.

Both alternators are connected to shunts and Alternator Current Sensors which show their individual contribution to the total charge. Measurements are displayed on the Monitor/Regulator.

13.7   The Battery Charger

The battery charger is not sized against the battery capacity. Some charger manufacturers indeed recommend that the charger size be correlated to the size of the batteries. In general, this recommendation is made to obscure a shortcoming in the charger. If we were to size the charger to the batteries, then the same rule applies to the charger as to the alternator. For conventional batteries, the charger would be rated at 25% of the battery *Ah*. For some sealed batteries the charger could be rated at 50% of *Ah* capacity.

While it would be nice to have a 450 Amp battery charger to match the batteries absorption rate, it isn't required. In fact, the power to drive such a charger would be about 8 kW, quite a bit more than we have available, even when tied up to the dock.

Since we don't expect to fast charge from the battery charger, its rating can be selected from other criteria. When tied up to the dock for long periods, batteries should be floated at a low voltage. To do so properly, requires that the charger be able to supply all the required load. The biggest load is the refrigeration system which requires about 40 Amps per compressor. By running only one unit at a time, the charger need only be a 40 Amp charger. And so it is. As mentioned in an earlier chapter, the 40 Amp charger is unregulated. It is used at dockside only when the refrigeration compressors run, but is used extensively offshore.

Supplementing the 40 Amp charger is a 10 Amp unit made by Schauer Manufacturing Company, Cincinnatti, Ohio. It is regulated with a Multi-Source Regulator. Charger output current is measured via a shunt and Charge Current Sensor which is coupled to the Monitor/Regulator.

13.8   The Solar Panels

All together on a sunny day, we have about 20 Amps of solar panel capacity. The solar panels often supply more than enough energy for our needs. To prevent overcharging, especially when the boat is left unattended, the solar panels are regulated by a Multi-Source Regulator with two All Purpose Drivers operating in parallel.

Solar panel current is measured with the same shunt and Charge Current Sensor used with the battery charger.

13.9   The Wind Generator

Our wind generator is strictly a home brew project. We even made the prop, balancing it statically with a couple of screws. A permanent magnet *DC* motor, which had a prior life as a machine tool servo motor, starts to generate in about 8 knots of wind. Peak output of about 12 Amps occurs between 15 and 20 knots. Peak vibration and noise also happens then.

We have a new *DC* motor which is yet to fly a prop. Preliminary tests indicate that it should be able to produce 20–25 Amps, provided that we build a big enough prop for it. The *DC* motor is a government surplus motor which undoubtedly cost dearly.

Wind generator current is measured with the same shunt and Charge Current Sensor used with the battery charger, and solar panel.

13.10   The Tow Generator

The tow generator is another home brew project. The first motor we had is now retired in favor of our new motor, mentioned above for the wind generator.

It is big and heavy, but produces 20 to 25 Amps at 7 to 8 knots. We use a 9 inch plastic outboard motor prop on the end of 50 feet of 1/2 inch double braid nylon to spin the motor.

The tow generator is not regulated, though it would be convenient to do so. Without regulation, the prop must sometimes be removed from the water.

Tow generator current is measured with the same shunt and Charge Current Sensor used with the battery charger, solar panel and wind generator.

13.11   The AC System

The *AC* system is comprised of two *DC* to *AC* inverters and one generator. One inverter is rated at 300 Watts, while the second is rated at 1000 Watts. Why two inverters? On the one hand, we wanted a big inverter that would run our bigger power tools such as the shop vacuum, 1/2 *HP* drill motor, and the small table saw. Most large inverters we looked at were power hogs, even without an *AC* load. The large inverters that had very little standby current drain, cost more than we wished to spend. By studying the *AC* load demands, we were able to divide them into two categories ... intermittent usage over an extended period of time, and heavy usage over a short period of time. For the appliances which were operated over extended time periods, the issue of standby efficiency was very important. For heavy, but short-term use, standby efficiency was not important. The computer, for instance, is used over an extended period. The toaster is only used for a very short period.

When we looked at the power requirements of the two lists, it became apparent that an inexpensive, but powerful inverter could be used for the heavy loads, while an efficient inverter could be used for the loads which were going to be used over a longer time period. Both inverters were purchased for less than the cost of a single efficient 1000 Watt inverter. The 300 Watt unit is more than sufficient to drive small tools, such as a drill or sabre saw. Because it consumes very little power when no AC load is on, we can leave the inverter turned on during minor repair jobs. When we need the 1000 Watt inverter, for instance to vacuum up after a repair, it is turned on only for the duration of the cleanup. Under heavy loads, the 1000 Watt inverter is about 85% efficient.

Both inverters are simple square wave types, although the 300 Watt unit was advertised as a "modified sine wave". We looked at its output on the oscilliscope one day out of curiosity. What we observed made us appreciate the quality power supply in our personal computer. That the computer can operate flawlessly on the AC generated by the inverter attests to the high design standard of the computer supply.

The AC Generator is a portable 650 Watt unit. As noted earlier, the battery charger is capable of delivering 40 Amps to the battery system at an output voltage of 14.4 Volts. The output power of the charger is thus 576 Watts. The charger is about 90% efficient, and so the load on the AC generator is about 640 Watts when the charger is operating at maximum. Since the charger is current limited at just over 40 Amps, the load on the generator is less when the batteries are discharged so that their output is less. For instance, at 13.8 Volts, the delivered power to the batteries is 552 Watts. This translates to 613 Watts from the generator.

Small generators are very inexpensive, and most of them are so quiet that they can't be heard for more than a few feet. They are lightweight and portable, and run for a fairly long time on a tank of fuel. And they can actually power the Balanced Energy System! A 650 Watt AC generator can drive the charger to its rated output,

40 Amps. That happens to be just enough to drive one refrigeration compressor. Forty Amps isn't enough to run the inverter to make toast, but that's where battery storage works. Once we sat in the dentist's office and overheard a conversation between a patient and a nurse. The nurse was building a new home, and had to pay an extra charge to get a wiring permit for 200 Amp service. We quickly multiplied 200 Amps times 115 Volts and came up with the shocking number of 23,000 Watts. We had a smile all the way to the dentist's chair.

The AC system is instrumented with both Volts and Amps sensors which isolate the AC system from the DC system, yet provide digital readings on the Monitor/Regulator. At dockside, it is very useful to monitor the current drawn so that the usual puny circuit breaker on the dock doesn't get overloaded. By the same token, it is nice to know how much voltage is available. When using our small generator, it is even more important to know what the voltage and current is. We like to run the battery charger at its maximum, while not overloading the generator. On hot days, the generator is given a little slack if the voltage starts to fall below normal.

13.12 Summary

On the one hand, the Balanced Energy System is a very personal system designed to meet our needs and wishes. On the other hand, it serves to illustrate the process which must occur for any performance energy system. The goal in designing the system was to optimize the generation, storage and utilization of energy. To do so meant that the components had to be treated as one system, and integrated accordingly. In the end, we compromised somewhat. We don't have enough solar or wind power to directly power the refrigeration system, so some stored battery power is required. On a sunny or windy day, the energy for the refrigeration system can be obtained, albeit at a slower rate than the refrigerator needs when running. We can power the refrigeration system from the battery charger and a small 650 Watt generator, with little to no wasted energy. Here at least, we have struck a perfect balance.

The main engine alternators are sufficient, though not as large as the battery system could handle. Restoring a near full charge to the sealed batteries is a relatively painless event, made easier by the fact that it can often be done while under way to a new lagoon.

The Balanced Energy System is well instrumented, so performance monitoring is easily accomplished. As such, excessive battery discharges are avoided, and the state of charge is known, preventing overcharges. Likewise, the instruments which measure refrigeration system performance allow precise control without wasted energy. Even the small generator is monitored via voltage and current measurements.

Chapter 14

Electrolysis and AC Safety

14.1   General Information

If you want to start a ferocious discussion among boating acquaintances just bring up the subject of electrolysis and *AC* grounding. Perhaps no other subject is cloaked with the mystery and mis-information of this subject. The real issue is safety and electrolysis. Can we be safe and also avoid electrolysis? With all the folklore surrounding this subject it comes as no surprise to see some of it printed from time to time in boating magazines as fact. It is possible to be safe without electrolysis problems, and this chapter is devoted to that subject.

14.2   Electrolysis Explained

As we know, a battery is constructed of two dissimilar metals which are placed in an electrolyte solution. Water is an electrolyte, and anytime two dissimiliar metals are submerged in it, a battery results. Electrolytic quality of water depends on the concentration of dissolved salts in the water. Saltwater is a much better electrolyte than fresh water. In fact the resistivity of saltwater is about 1/70th of the resistance of fresh water. If you dangle a power cable in saltwater, 70 times as much current will flow as would flow in fresh water. Electrolysis then, is mostly a problem in saltwater. As we will see later, safety is a bigger problem in fresh water.

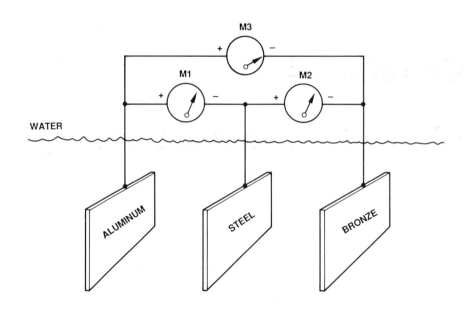

Figure 14.1: The Water Battery

Very few metals are inert in saltwater. Some will dissolve readily while others can stand total submersion on a continuous basis. The more inert of the metals are called more *noble*. Less noble metals migrate to more noble metal as electrical current flows between them.

In Figure 14.1, three plates of metal are shown submerged in salt-water. The plates represent the underwater surface of boats . . . perhaps the hull, or machinery such as prop shafts, *I/O* units and rudders. Aluminum, being the less *noble* of the metals forms a positive plate with both the steel and the bronze. Steel is less noble than bronze, and thus forms a positive plate with it, even though the steel is the negative plate for the aluminum/steel battery.

The meters shown in Figure 14.1 indicate the direction of current flow through the water battery itself. Through a battery, electron flow is from positive to negative, just the opposite of electron flow through a load.

Galvanic action depends on factors other than just metal types. Some lesser grades of stainless steel corrode much faster than might be expected when is submerged in water and deprived of oxygen. We have experienced no problem with our underwater stainless which is type 316. On the other hand, our lifeline stanchions which are of an unknown type seem to rust on an overcast day. As a friend pointed out once, stainless steel is only guaranteed to stain less.

Itemized below is the relative nobility of various metals. The least noble metals always dissolve before the more noble metals listed at the top.

Nobility Series of Various Metals

Platinum

Gold

Silver

Inconel

Stainless 316

Stainless 303

Monel

Nickel

Cupronickel

Silicon Bronze

Copper

Brass

Tin

Lead

Cast Iron

Wrought Iron

Mild Steel

- Aluminum (6061, 1100)

- Galvanized Steel

- Zinc

- Magnesium

   The current flow between the plates in the water battery can be likened to battery self-discharge. As the self-discharge current flows, the aluminum will rapidly dissolve. Each electron leaving the aluminum for the steel and bronze will carry away a metal ion. Given enough time, the aluminum will completely disappear. Likewise, the steel will dissolve as electrons leave it for the bronze. Eventually the bronze will be the sole survivor.

   Strictly speaking, electrolysis cannot be prevented between dissimilar metals, but its damaging effects can be prevented by sacrificing a soft metal, usually zinc. Two steps must be taken. First, all underwater metal must be connected internal to the boat with heavy wire called a *bond*. (More on bonding techniques later.) Secondly, zincs are mechanically and electrically attached to underwater metal.

   Figure 14.2 shows the metal plates with zincs attached. Note that the current between them, shown by *M1* and *M2* are greatly reduced. The zinc forms a positive plate with all the plates, and since the zinc is electrically attached to each plate, no overall battery results. Note that on the other meters, *M2* and *M4*, some electrical activity is shown. The metal which is not close to the zincs still acts as a plate in a battery. Moral: Distribute your zincs to protect both fore and aft.

   Again, no connection exists between the boats in Figure 14.2. If each boat has adequate zincs then no damaging electrolysis will occur. So wherein lies the problem?

14.3   The Little Green Wire

The green wire in the *AC* power cord carries no current under normal operation. *AC* appliances work as well without the green wire as they do when it is connected. We all know about the adapters which are

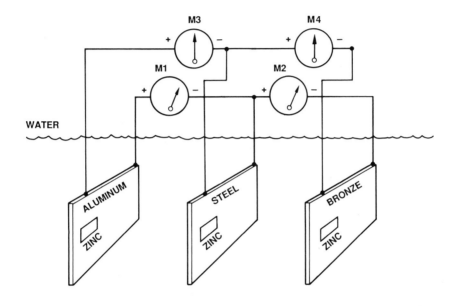

Figure 14.2: Zinc Protection

used when a three prong plug must be connected to a two wire outlet. Though it appears unnecessary, the little green wire is truly a *life and death* situation.

The green wire at the power generation plant is connected to earth by a long metal post which is driven deep into moist soil. Its purpose is to provide a return path for loose, or stray currents. These stray currents result when leakages occur within AC appliances. A typical cause of leakage might be a mica insulator in a marine battery charger which has been punctured or improperly assembled[1]. The case of the battery charger is normally connected to the green wire, and if leakage occurs, the stray current can return to earth safely. Without the green wire, the leakage may be conducted through a person ... death may result!

To evaluate the potential for a deadly shock, two questions must be considered. How much current does it take to kill, and what is the possibility of exposure to that current? Exposure to electrocution falls in three classes ... inside the boat, contact between boat and dock, and swimming in the near vicinity.

An ungrounded appliance with leakage current, such as our battery charger example above, represents the most serious threat to people on the boat. In a wooden or plastic boat, you might be able to touch a *hot* appliance without getting a shock because the surface that you are standing on is probably insulated. An ungrounded appliance on an aluminum or steel boat is just an electric chair waiting for a victim. To prevent killing occupants of your boat, you must have a good return connection to earth, or you must use an isolation transformer which is described later.

The second form of exposure to deadly electrocution occurs when a person on the dock grabs a metal part on your boat. If your boat is not grounded, that person becomes a human conductor. This is illustrated by Figure 14.3. Again, a good ground connection is required to prevent killing those who touch your boat. Leakage current from shore power must find a return path to earth ground. A person standing on a grounded dock who touches your boat may be the first to learn of the short circuit.

---

[1]We diagnosed this problem for a Seattle firm which sells cruising equipment when the owner was unable to find anything wrong with the battery charger.

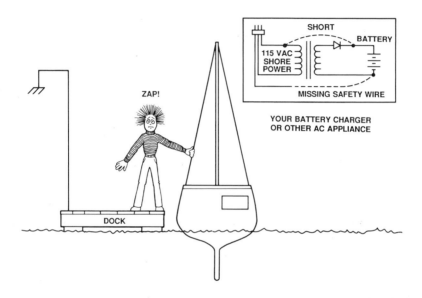

Figure 14.3: Killing Someone You Love

A third form of shock exposure results when people swim nearby a boat which is ungrounded and has an internal short. Studies by UL[2] have shown that 2 Volts *RMS* can be deadly to swimmers in the vicinity of an ungrounded boat. This small *AC* voltage is able to conduct 5 milliamperes of current through the human body ...more than enough to be lethal!

The surest way to prevent exposure to electrocution is to have a good connection between your *DC* electrical system, all metal parts on your boat, and the *AC* green safety wire. It's a simple thing to do, so why all the hassle? The fact of the matter is, connection to the green safety wire can/will increase the odds that you suffer from electrolysis.

14.4 Electrolysis Re-visited

In Figure 14.4, the metal parts of four boats are shown. Three of the boats are connected to the *AC* safety wire. One of them is isolated from the connection. The additional connections between the boats means a greater risk of electrolysis. The green wire is often poorly connected to the driven earth post. Resistance in the wire, and poor terminations means that the boats at the end of the pier have a bad connection to ground. In fact, in saltwater, contact with the water itself is a better connection to earth than the green wire. The resistors in the wire are the result of poor wiring connections and long wires along dock fingers. The aluminum boat continues to corrode away to the steel boat.

The poorly connected green wire, and the long wire lengths from one end of the dock to the other, guarantees that an impressed current will flow between the connected boats. As shown in Figure 14.4, the aluminum metal will fast disappear. Even the boat with no green wire connected will be a plate in this oceanic battery, but it is only connected via the electrolyte and is not driven by the voltages developed along the length of the green wire. In sea water, it turns out that electrolysis problems can be lessened by *not* connecting to the *AC* safety wire. What about exposure to electrocution? In Table 14.1

---

[2]Underwriters Laboratory.

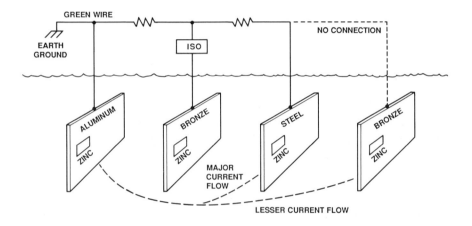

Figure 14.4: Connected Boats

| | Fresh Water | Sea Water |
| | Connected/Not | Connected/Not |
Risk Type		
Electrocution Inside Boat	Low/High	Low/Low
Electrocution Outside Contact	Low/High	Low/Low
Electrocution Swimming Near	Low/High	Low/Low
Electrolysis	Low/Low	High/Low

Table 14.1: Consequences of Green Safety Wire Connection

safety and electrolysis risks are presented.

Based on the summary presented in Table 14.1, it appears that boats in fresh water should be connected to the AC safety wire, while boats in sea water should not be connected. Standards of the National Fire Protection Association and the American Boat and Yacht Council do not permit sea based boats to be treated differently than fresh water boats. This is done for good reason. Boats are mobile, sea water today, fresh tomorrow. Even the salt content in sea water is not guaranteed to provide the necessary low resistance to earth. At many marinas, enough fresh water flows into them to cover the surface of the saltwater and thus insulate your boat from the earth ground. Only a fool would recommend that you ignore the mandate of authorities and not connect your DC ground to the AC safety wire. Without an isolation transformer or an *isolator*, direct connection is a must.

14.5   Prevention of Electrolysis

The surface area of exposed metal determines the amount of current that will flow between the dissimilar metals, while the type of metals determine the amount of voltage generated. Prevention of electrolysis can be done in several ways. First and foremost, you must bond all underwater metal together and to your DC ground.

By bonding, we mean to connect all points with a heavy copper conductor. Solid AWG #6 or AWG #8 copper wire is sold at electrical distributors, and is appropriate for bonding. Better yet, buy some copper strap that is at least 1/2 inch wide and 1/32 inch thick. Run the strap along the inside hull on either side, and after attaching all the metal parts of the boat to it, fiber glass over the strap to prevent corrosion. Your electronics, especially your *HAM* or *SSB* radio will work all the better for the good ground plane provided by the copper strap.

Depth sounder transducers, water intake fittings, stern tubes, zinc throughbolts, prop shafts and all other underwater fittings must be connected. While you are at it, connect the chain plates to provide lightning protection. The bonds must make good electrical connection. Start by cleaning the fittings with a stainless steel wire brush or bronze wool. If you cannot solder the copper to the fittings, then wrap several turns of wire around the fitting. Now apply a small amount of petroleum jelly and then put a stainless steel hose clamp around the wire and the fitting. Attach it firmly.

Bonding prop shafts is not always easy. If you are using a flexible coupling, then put a bonding wire between the two halves. This will at least connect the shaft to the transmission. You should measure the resistance from the prop shaft to the bonding system attached to the engine. If it is more than several tenths of an ohm, then you should connect up a shaft brush assembly.

Don't forget to bond your stern tube and the packing gland housing. The latter is often attached to the stern tube with a rubber hose and is thus insulated from the stern tube. We forgot the packing gland on our own boat until we noticed some telltale purple/black pocks.

You should note that bonding wires are not to be used as current carrying conductors in the electrical system. To do so will hasten any electrolysis. Bring all bond wires to your central negative distribution terminal. Don't use an engine block as a connection point, but be sure your engine is connected to the negative distribution terminal with a heavy conductor.

Electrolysis can also be prevented by an impressed *cathodic current*. A special hull fitting is driven positive, or *anodic* with respect to all underwater (cathodic) hardware. This protects the cathodic metal. The owner of the impressed current system protects his/her boat as well as other surrounding boats. The more boats there are in the surrounding area, the more will be the impressed current. This

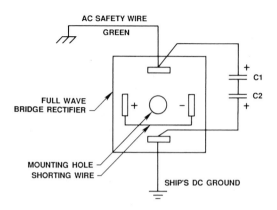

Figure 14.5: Diode Bridge and Capacitor Isolator

isn't such a good deal for the owner, who may find batteries going flat as a result. If the system doesn't automatically adjust the current, then as the boat population changes, the impressed current will be too much or too little. All things considered, we don't recommend an impressed current system except for very *hot* marinas with no other options.

There are two ways that you can isolate your electrical system from the *AC* safety wire and still enjoy the safety which the wire provides. The first of these is shown in Figure 14.5. By using a standard diode bridge connected as shown, you can avoid a direct connection between *AC* earth ground and your *DC* ground. Instead of the direct connection, there will be two diodes which separate the *AC* from the *DC*. This will allow up to 1.5 Volts to exist between the two grounds. Electrolysis currents will not flow, but should a voltage larger than 1.5 Volts occur, then the diodes will conduct. Also shown are two electrolytic capacitors which are connected between the *AC* and *DC* grounds.

The diode bridges have a convenient mounting hole and an aluminum base plate which is isolated from the diodes. Be sure to short the two *DC* connections labelled (+) and (-), as shown. Select a diode bridge which is rated for about 5000 Amps for 20–30 milliseconds. This will allow enough time for a magnetic/hydraulic circuit breaker to open. If one diode bridge won't do, use several in parallel.

The two electrolytic capacitors are connected *back-to-back*, that is their negative terminals are connected together. This results in a non-polarized capacitor. The final capacitance value should be on the order of 2 Farads. To get that value, you will need two capacitors at 4 Farads each. A large voltage rating is not required because the diodes limit the maximum voltage to which the capacitors are exposed. (You can buy 1 Farad capacitors at 6.3 Volts. You will need 8 of them to end up with 2 Farads, non-polarized.)

Because you will need to parallel several diode bridges, and use the large capacitors indicated, the cost to build the isolator may be around $75. You may want to invest a little more and get an isolation transformer instead.

Aluminum or steel boats should be equipped with an isolation transformer as shown in Figure 14.6. Note that the *AC* safety wire is connected to the transformer case, but does not continue any further. The *AC* used inside the boat is from the transformer secondary. No current flows between the primary and secondary windings, so any shorts on the secondary do not seek an earth ground return path.

Isolation transformers can serve two purposes. First, they can reduce the risk of your own appliances developing leakage current which returns to the *AC* ground via your zincs, eating them up in the process. The second reason you might chose to install an isolation transformer is for voltage conversion reasons. With the proper transformer, you can visit Europe and be assured that you can accept 220 *VAC* instead of the 115 *VAC* used here in the U.S. For this you will need a dual primary transformer that can be switched between 115 *VAC* and 220 *VAC*.

If you go to the expense of installing an isolation transformer, we recommend that you get one with a full *box* shield around the primary. The box is a 100% metal enclosure around the primary that precludes any possibility of leakage between primary and secondary. Usually the box is contructed of light gauge brass which does not interfere with the usual magnetic action in the transformer. Such transformers are required in all medical instruments which can be attached to patients.

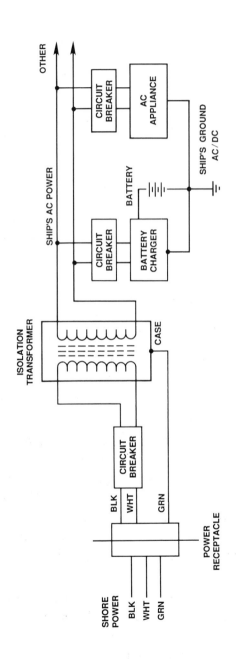

Figure 14.6: Using an Isolation Transformer

14.6 Electrolysis Sidelight

In Puerto Vallarta, Mexico, we had the bad experience of losing all the galvanized plating on the submerged portion of our anchor chain. At the time there was a sunken metal boat in the harbor. It was a huge unprotected metal plate which we anchored next to. Fortunately, we had sufficient zincs protecting our prop, prop shaft and rudder shaft to prevent electrolysis of those parts. Only a few days were required to strip all the galvanizing from the chain. We had to swap our chain end for end when we left, wiser about the wily ways of electrolysis.

14.7 Summary

Safety is paramount, but a direct connection of the AC green wire to your electrical system will lead to electrolysis problems. Two ways of enjoying safety while avoiding electrolysis are the isolation transformer or the diode/capacitor isolator. Don't take unnecessary chances with the lives of others ... either use an isolation transformer, a diode/capacitor isolator, or make a direct connection to the AC safety wire!

Chapter 15

Miscellaneous Topics

15.1   General Information

This chapter contains topics of interest to persons selecting or wiring equipment for an alternate energy system.

15.2   Displays ... *LCD* vs *LED*

There are two principle types of displays used for small instruments. The type which glows red is the light emitting diode (*LED*) type. Those which are usually grey, are liquid crystal displays (*LCD*).

The *LED* display is easily read in the dark, but usually washes out in direct sunlight. If its contrast is high enough for daylight viewing, then after dark, the display is obnoxiously bright.

The *LCD* display, on the other hand, is easily read in daylight, but requires some other light source after dark. The *LCD* display also has to be viewed more perpendicular to its face for readability, while the *LED* display can be read from more oblique angles.

Given these parameters alone, the choice between the two types of display would be a tough one. There are other aspects about the displays, however, which make the choice easier.

The *LED* display is a power hog compared to the miniscule draw of the *LCD* type. In cases where power is plentiful, the *LED* display can be read in total darkness ... it helps to have a brightness control.

If the display never needs to be read in direct sunlight, the *LED* would offer a distinct advantage. Other factors must be considered, however. Ample Power Company elected to use an *LCD* display in the Monitor/Regulator. This decision was based on the following reasoning:

- Power consumption must be held to a minimum, allowing continuous operation of the *M/R* without undue battery load. This is particularly important when the *M/R* is used to measure the *at-rest* voltage of a battery, since larger currents remove the battery from its rested state and thus destroy the accuracy of the reading.

- The *M/R* would be used in direct sunlight, but probable never in total darkness.

- An *LCD* display would allow the *M/R* to provide excellent accuracy of measurements. This latter point is a subtle point that even many engineers might miss.

Most small digital meters use one of a series of integrated circuits made by several circuit manufacturers. GE–Intersil[1] is one of the better known manufacturers, and has a reputation for innovation and quality. It produces 4 integrated circuits (chips) of interest ... the 7106, 7107, 7126, and the 7136. The 7107 is designed to drive *LED* displays, while the 7106, 7126 and 7136 are designed to drive *LCD* displays. The 7136 is a lower power version of the 7126, which is a lower power version of the 7106. Technical literature from Intersil clearly states that the 7107 *LED* chip is unstable if the internal drivers are used with the internal reference. The internal heating on the reference varies according to how many segments are on inside the *LED* display. (Each digit in a display is made up of 2 to 7 segments.) Under the circumstances, it is impossible to get an accurate reading unless an external reference is used, and even then the internal chip heating and the current drawn by the *LED* drivers makes accuracy improbable. As the digits flicker, different amounts of current are drawn, and each discrete value causes the next conversion to be different. The result is a continual change of several digits. A manufacturer may try to hide this by using a low rate of conversion, ... one reading every 3 seconds.

---

[1] Intersil, Inc., 10600 Ridgeview Court, Cupertino, CA 95014

n use, this slow a rate is aggravating, since it forces the user to wait or a reading after a channel has been selected.

Ample Power Company uses the lowest power, and highest priced of all the chips, the 7136. As a result, the Monitor/Regulator is accurate within seconds after the application of power, and it stays accurate over time and temperature. Even at 3 conversions per second, the M/R exhibits a low rate of one digit flicker. The M/R draws a mere 0.009 Amps, which doesn't change with the reading. A single segment in an LED draws this amount. A reading of 13.88 Volts (22 segments) on an LED display will draw about 0.200 Amps ... 22 times the amount that the M/R draws!

Obviously, 0.2 Amps drawn from a battery will mean that the battery is not in a rest condition, so the principle reason for accurate voltage measurement is aborted. Secondly, 0.2 Amps current through the voltage sensing leads will also cause voltage drops so that what the meter reads will not be the same as the voltage at the battery, unless separate power leads are used. Should you forget to turn the LED display off after using it, the continual draw of 0.2 Amps will eventually discharge your batteries.

For battery voltage instrumentation, the choice seems clear. If all you want is numbers which glow in the dark, then the LED is well suited. If you want those numbers to mean something, the LCD is the way to go.

15.3 Accuracy and Resolution

To the unsuspecting, a reading of 12.20 Volts would seem to indicate that the battery voltage is somewhere between 12.19 and 12.21 Volts. *Alas, it ain't necessarily so.* From our first digital voltmeter design in the early 1960's, to the design of the Monitor/Regulator in 1986, the issue of resolution and accuracy has been of paramount concern.

Resolution is easier to explain than accuracy. A reading of two decimal places, for example 12.20 Volts, has a resolution of 10 millivolts. That is to say, 1 digit on the display represents 10 millivolts (0.01 volts).

Accuracy is quite another matter altogether. To begin with, any digital reading is subject to at least a 1/2 digit error. Suppose for instance that the true voltage is 12.195. By showing 12.20, the reading is off by 0.005 Volts (5 millivolts). In general, the point at which a meter switches from one digit to another causes a full digit of error minimum. At full scale, one digit may not be a significant error. A low scale, say 5 digits, a one digit error is a whopping 20%.

Besides the built-in one digit of error, there is also an error which is a percentage of the reading. This error results from the time and temperature drift of the internal reference voltage which is used to compare the input voltage. All the components in the reference circuit are subject to time and temperature drift, including the potentiometer used to initially calibrate the instrument. Errors can also occur because the analog to digital conversion is not exactly linear Other types of errors occur, such as common mode errors. Yet other errors occur because of internal leakage currents. Usually, all errors are lumped together in a worst case number, and accuracy is specified as X% of reading plus or minus Y digits.

To make this more sensible, consider a specification that says the reading is accurate to 0.1% of the reading plus or minus 1 digit. Full scale on the Monitor/Regulator is 20 Volts (19.99). 0.1% of 20 Volts is 0.02 Volts. Adding the one digit to this error results in an error of 0.03 Volts maximum, or three digits total. This is, in fact, the typical accuracy of voltage measurements of the M/R at 77°F. Over temperature extremes to 125°F, the reading can be in error as much as 0.3% of reading plus or minus 1 digit for voltage measurements.

Additional factors must be taken into account to determine the accuracy of current measurements. To minimize internal errors due to leakage currents in the input channel multiplexer, the analog to digital chip in the Monitor/Regulator is operated at 2.000 Volts full scale. This means, however, that amplifiers are required to measure currents. For example, shunts which develop 0.050 Volts (50 milli volts) are used to sense current. To convert 50 mV to 2 Volts requires amplification by exactly 40 times. Amplifiers of course, can contribute error. Furthermore, the resolution of the M/R is 1 part in 2000, or 0.0005. Take 0.0005 times 0.050 Volts and you arrive at the rather low value of 0.000025 (25 micro-volts). When measuring current then, 25 uV represents one digit on the M/R. Needless to say, it takes a rather special amplifier to measure 25 uV.

For battery current measurements, Ample Power Company uses a special temperature stabilized amplifier that has an initial offset of less than 10 uV, with a temperature drift of 0.05 uV per $°F$. With this kind of initial accuracy and low drift, current measurement accuracy will be totally dependent on the shunt which is used to sense current. The shunts have an initial accuracy of 0.25%. It is safe to assume that battery current measurements are accurate to 0.35% of the reading plus or minus 1 digit. Note that some instrumentation manufacturers use shunts with an initial accuracy of only 2%.

Measurement of alternator current is slightly more difficult because the shunt is connected in series with the positive output. This results in both inputs to the sensing amplifier at a high elevation from ground potential. Ordinary techniques under these conditions would yield highly inaccurate readings. Ample Power Company uses the best possible amplifier for the alternator current sensor. Initial offset is trimmed to zero, and offset drift is about one digit for every $10°F$. Alternator current measurements will thus be within 0.35% of reading plus or minus 6 digits.

Some other instrument manufacturers operate their analog to digital chip at 200 mVolts full scale instead of 2 Volts ... they suffer from multiplexer leakage currents as a consequence. If they use mechanical switches instead of solid state multiplexers at 200 mV, then they suffer from electrical contact noise and thermal offset voltages. These same manufacturers use no sensing amplifers next to the shunts ... they suffer great noise and temperature problems measuring current, particularly if there is a long wiring run from the shunt to the measurement instrument. By taking these shortcuts, a system can be built cheaper than one which uses external sense amplifiers mounted near the shunts. The *resolution* of a cheap system may be the same as that on a more expensive one, but *accuracy* will not be as good. All too often, however, the user doesn't know the difference between resolution and accuracy. Ignorance may be bliss, but ignorance rarely leads to peak performance.

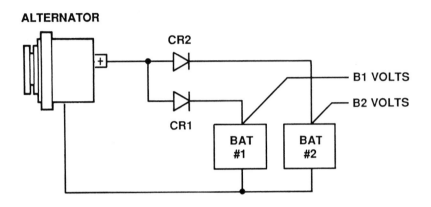

Figure 15.1: Alternator with Isolator Diodes

15.4   Regulation and Isolator Diodes

Isolation diodes are used in three principal places ...as isolation diodes between the alternator and the batteries, within multiple output battery chargers, and in series with solar panels and wind generators. The use of diodes in these three places serves the same purpose, to prevent current flow out of the batteries, while allowing multiple batteries to be charged simultaneously. A problem exists with diodes used for isolation, however. While the problem is the same for alternators, battery chargers and solar or wind chargers, the problem will be illustrated here with the common alternator.

Figure 15.1 shows an alternator with two isolation diodes, CR1, and CR2. A dual output battery charger uses two diodes in a similiar circuit as that shown in Figure 15.1.

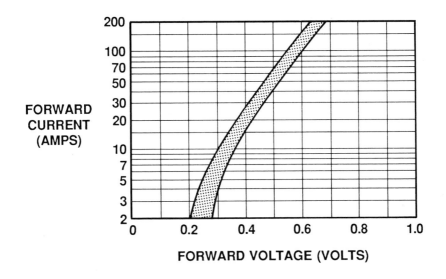

Figure 15.2: Forward Voltage Drop vs Current

Figure 15.2 shows the forward voltage drop versus current for a specific Schottky diode[2] used as a 200 Amp Fail Safe Diode. Diodes are definitely not all the same, but silicon diodes in general have more voltage drop than do Schottky diodes. Note in Figure 15.2 that at about 2 Amps, the diode drops about 0.2 Volts. At 150 Amps, the forward voltage drop has increased to 0.6 Volts. Of course the values plotted in Figure 15.2 are typical values ... some diodes will be better, and some worse. In a high quality Schottky diode, such as the 200 Amp Fail Safe Diode, the voltage drop from one diode to the next will be close to the same. The major problem in using diodes results from the difference in voltage drop with current.

To see how this difference in voltage drop causes a problem, some

---

[2] Ample Power Company, # 1029.

background information about multiple battery voltage sensing regulation is necessary. A regulator that senses voltage from multiple batteries actually uses the highest voltage as the regulation feedback. Referring to Figure 15.1, assume that Battery #1 is fully charged, while Battery #2 is fully discharged. When the alternator is first turned on, Battery #2 will demand most, if not all of the current from the alternator. This will *drag* the alternator output voltage down, and little or no current will be able to flow into Battery #1 because diode CR1 will be reverse biased.

Eventually, the voltage on Battery #2 starts to rise, as the charge builds up. At some point, the voltage will rise high enough so that a small current starts to flow into Battery #1. The regulator, sensing that the highest battery is up to the regulation voltage, will hold that setpoint. Assume for a moment, that the alternator is being regulated at 14.4 Volts, as sensed on Battery #1. Battery #2, however, can still absorb a lot of current. Perhaps, Battery #2 is a house battery, and as it is being charged, it is also being loaded. Now, Battery #1 needs no charge, and so the current through CR2 is very small, say 2 Amps. On the other hand, Battery #2 is absorbing a high current, plus it is also loaded. That means the CR2 is conducting a much higher current than CR1, and as a result is dropping a higher voltage. While Battery #1 voltage may be at 14.4 Volts, the voltage on Battery #2 may only be 14.0 Volts. Under the circumstances, Battery #2 will not fully charge as fast as it might if its voltage were actually 14.4 Volts.

The above scenario is typical of many *RV* systems, and to a lesser degree typical of many boat installations. Generally, an *RV* has a single starter battery and a single house battery. With a dual voltage sensing regulator, the house battery which gets discharged the most, and is usually charged with a load attached, doesn't ever get fully charged!

The problem can be alleviated, if not totally solved by several means. The easiest solution is to simply remove the voltage sensing from the starter battery so that the regulator will see only the voltage of the house battery. This simple approach has a bad side effect ... the starter battery will see elevated voltages in the vicinity of 15 Volts. Obviously the starter battery will have a short life. A second solution to the problem is to simply connect the two batteries together during charge. This means that the diodes have no purpose other than as a fail safe mechanism in the event of an alternator short circuit.

A third, and better alternative, is to employ a two battery system where both batteries are the same Ah capacity, and are used on alternate days. Then when charging occurs, both batteries will be discharged, and the currents through CR1 and CR2 will be more or less equal. If, however, one battery is loaded during charge, then there will still be a difference in the two diode currents, and the loaded battery will still not get a full charge.

Isolators don't perfectly solve the problem of which battery to charge. Used improperly, as is the typical case, they create problems rather than solve them. If used in the two equal battery system, with no load on either battery during charge, then the two batteries will charge more or less equally. We recommend at least one Fail Safe Diode in series with the alternator as insurance in case of an alternator short. With that single diode, we recommend a selector switch that directs the charge to either or both of the batteries.

For the single starter battery and single house battery system, the connections shown in Figure 15.3 are appropriate, although the Fail Safe advantage is lost. As shown, the starter battery is fed via a diode, while the house battery is connected directly to the alternator. The diode must be a Schottky type with low forward voltage drop. Now, the house battery voltage will dominate the regulator, while the starter battery will trail by 0.2 to 0.3 Volts. This would be a problem if a single setpoint regulator is used which is set at 13.8 Volts, since the starter battery will never fully charge. With a regulator such as the Automatic 3–Step Deep Cycle Regulator[3] which has a high absorption voltage, then the starter battery will receive adequate charge. For instance, if the house battery is charged to 14.4 Volts, then the starter battery will get to 14.1 to 14.2 Volts. This is quite sufficient for a starter battery, particularly if it is a maintenance free unit with lead calcium plates since they charge at a slightly lower voltage.

---

[3]Ample Power Company Seattle, WA

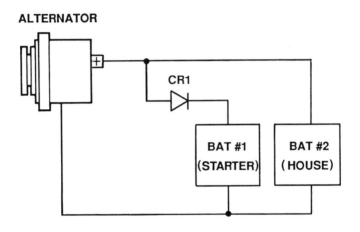

Figure 15.3: Using a single diode for Isolation

15.5 Charge/Access Diodes

The diode is a vital part of most electronic equipment. A simple but important example of diode usage is the Charge/Access Diode. Diodes, which are often called rectifiers, allow the passage of current in only one direction. When a diode is connected in a forward direction, the diode drops a small voltage and passes current readily. A diode connected in the reverse direction blocks the flow of current. Some small leakage current does flow in the reverse direction, but too little to be of concern except in sensitive circuits.

Diodes have several important design parameters. They are:

- the amount of forward current that can be conducted.

- the amount of forward voltage drop at the applied current.

- the maximum reverse voltage that can be applied before catastrophic diode failure.

- the thermal resistance of the diode junction to the diode case. This latter parameter is necessary to calculate the surface area needed to properly heatsink the diode.

Diodes are made in current ratings from milliamperes to thousands of Amps. A Schottky diode under useful load current exhibits a forward voltage drop of 0.5 to 0.8 Volts. The voltage drop decreases with higher temperature. Expect 0.65 Volts under most loaded conditions. The maximum reverse voltage which can be applied to a Schottky diode is generally 15–20 Volts with more expensive diodes limited to 35–45 Volts. Thermal resistance varies according to the diode current rating and falls between several °C per watt dissipated to a fraction of a degree per watt.

Schottky diodes are preferred over silicon diodes because of their low forward voltage drop. A silicon diode at useful load currents drops from 1 to 1.5 Volts, with 1.2 Volts being typical. Note that a Schottky diode which drops 0.65 Volts from a 14 Volts source, such as a solar panel, has lost almost 5% of the available voltage.

Figure 15.4 shows the Ample Power Access Diode[4]. It consists of a dual Schottky diode mounted on a heat sink and attached to a

---

[4] Ample Power Company #1034.

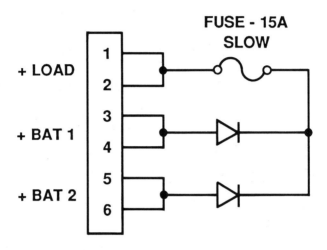

Figure 15.4: Charge/Access Diode

printed circuit board. A replaceable fuse is connected as shown, to the common diode cathodes. Connection to the dual diode is made via a six pin terminal block. Two pins each are used, allowing ease of hook up. When pins 3–4 or pins 5–6 are positive with respect to pins 1–2, then current will flow. The dual diode and circuit board assembly is potted in epoxy to prevent moisture penetration.

The fuse on the assembly is a 15 Amp *slow blow* type, which is sufficient for the rated load of 12 Amps. The dual diode has been selected so that the maximum load current can flow through either of the diodes, or both of them in parallel. Maximum reverse voltage is 35 Volts, with reverse current less than 15 milliamperes. Typically reverse current will be less than 1 milliampere. The heatsink has sufficient surface area to keep the diode junction within rated temperature at an ambient temperature of 140°F. At rated load of 12 Amps, the

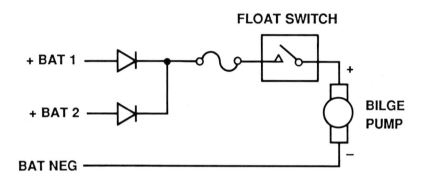

Figure 15.5: Access Diode with Bilge Pump

heatsink will operate about 90°F above ambient. The Access Diode is supplied with nylon standoffs which are used to thermally isolate the heatsink from its mounting surface, if used under continuous full rated load conditions.

Figure 15.5 shows the Access Diode connected to a bilge pump. In this application, the pump has access to the available energy of both batteries. Both batteries must be dead before the bilge pump quits.

Figure 15.6 shows the Access Diode connected to electronics. In particular, this configuration keeps power applied to the equipment even if one of the battery voltages is abnormally low. For instance, when an engine is started on one of the batteries, the voltage of that battery may fall as low as 6 Volts momentarily. Many Lorans will dump their stored waypoints at that voltage. With the Access Diode connected, the battery with the highest voltage supplies the power for

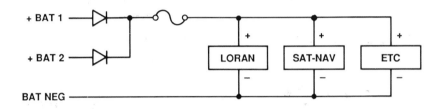

Figure 15.6: Access Diode with Electronics

the electronics, thus the battery which is not used for starting saves the Loran waypoints.

Figure 15.7 is the same as Figure 15.6 with the addition of an electrolytic capacitor. A capacitor stores energy in much the same fashion as a battery. By using the capacitor with the Access Diode, both batteries can be used for starting. The energy stored in the capacitor is prevented from flowing back to the batteries by the diodes, thus is available for the electronics. A capacitor value of 20,000 microfarads is sufficient for low power loads such as a single Loran unit. Additional capacitance would be required for more loads.

Figure 15.8 shows the Access Diode used in conjunction with a complex filter. Such a filter contains inductance to prevent noise passage, and includes a capacitor bank to both filter and provide energy storage so that temporary loss of power does not cause electronic

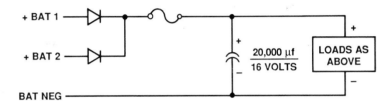

Figure 15.7: Access Diode with a Simple Filter

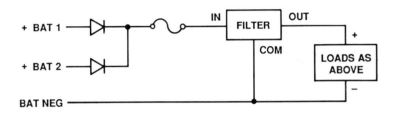

Figure 15.8: Access Diode with a Complex Filter

memory dump.

Figure 15.9 demonstrates how the Charge/Access Diode is used in conjunction with wind generators or solar panels to charge two batteries. It should be noted that both the wind generator and solar panel will draw current from the battery when they are not operating unless a diode is used to prevent reverse current. The Access Diode prevents reverse current, and by using two of them, the charge is split to the batteries. Two Charge/Access Diodes are not necessary if the output is wired to the battery selector common terminal. Note that both diodes in each assembly are operated in parallel when used as a Charge Diode. Because the diode itself is a dual diode assembly and closely matched, the current flow is shared more or less equally. Thus at the rated flow of 12 Amps, each diode will conduct about 6 Amps. At this lower current, the forward voltage drop of the diode will be

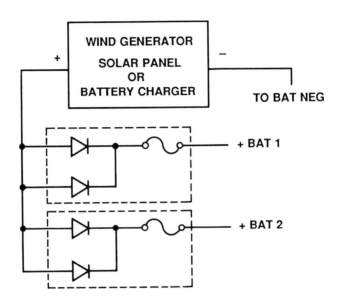

Figure 15.9: Charge Diode Application

on the order of about 0.4 Volts ... less precious charging energy is lost to the diode.

The Charge/Access Diode can also be used with small battery chargers as a splitter, allowing two batteries to be charged independently. At high initial currents, there will be a forward voltage drop of about 0.65 Volts. As the batteries reach a full charge, and current falls below an Amp, the forward drop will fall to about 0.3 Volts. For automotive type battery chargers, this drop may be just enough to keep from boiling water out of the batteries. Measure the voltage once the battery is fully charged to verify that the voltage is not greater than about 13.8 Volts.

While 12 Amps is sufficient for the majority of applications, Access

Diodes can be operated in parallel for higher demands. Diodes in different assemblies may not share current equally, however. When parallel assemblies are used, derate the sum of the individual ratings to 75%. For example, four units can be used in parallel to conduct 36 Amps, not 48.

The Charge/Access Diode is a valuable component in the electrical system. The efficient dual Schottky diode was chosen against a carefully considered set of parameters, and a proper heatsink was designed to maintain junction temperature within rated limits. The whole assembly is fuse protected, and potted in epoxy to assure many years of trouble free operation.

You may elect to make your own Access Diode. Besides selecting a diode which can pass the necessary current, it is important to provide enough heatsink surface area to dissipate the power consumed by the diode. Note that a diode must be derated when temperature increases. While a diode may be rated at 30 Amps, at $25°C$, as temperature rises to $100°C$, (boiling) the diode may only be used at 5 Amps. It pays to keep the diode as close to ambient temperature as possible, but the price paid is a large heatsink.

Diodes are specified with three important details which allow calculating the necessary heatsink surface area. For instance, a diode may be rated at 30 Amps, with a thermal resistance of $1.5°C$ per Watt of power dissipation. Additionally, a maximum junction temperature of $150°C$ is specified. Suppose we wish to use the diode at 20 Amps, and at that current, the diode drops 0.6 Volts. The diode dissipates 12 Watts (20 times 0.6). At $1.5°C$ per Watt dissipated, the junction temperature will rise $18°C$ above its mounting base temperature. This sets the maximum heatsink temperature at $132°C$ (150–18). This temperature is above boiling, and while the diode may be able to stand the heat, the surrounding objects may not be able to. Assume we wish to keep the heatsink temperature at or below $80°C$. How much heatsink surface area is necessary?

This problem is similar to the heat conduction problems associated with refrigeration evaporators and condensers. For those problems, we used a rule of thumb that 2.5 *Btu* per square foot per degree per hour is conducted from a metal object into still air. This is a conservative rule of thumb for refrigeration work. For electronic work, where size is generally more critical, and heatsinks are usually black to increase their effectiveness, a second rule of thumb can be used. Again, the rule is based on surface area, as well as the temperature difference between the heatsink and the ambient air.

Assume that we allow for an ambient of $50°C$ (about $120°F$). If we want to keep the heatsink at $80°C$ or less, then a rise of $30°C$ can be tolerated.

To calculate the surface area required, plug into the rule of thumb given below.

$$SA = (125)(W)/Tr$$

In this rule, SA stands for surface area, measured in square inches, W stands for Watts, and Tr stands for temperature rise, measured in $°C$.

For our example, the surface area required is 50 square inches. ($50 = (125)(12)/30$). If you want to get this surface area in a small size, then a heatsink with multiple fins should be used, otherwise, the 50 square inches can be obtained by using a 5 inch square piece of aluminum which has 25 square inches of surface area on each side. For good heat conduction through the aluminum, a thickness of 0.06 to 0.09 inches is recommended.

15.6 Noise Filters

Electrical noise is a phenomena which is exhibited anytime that current flows. Even a simple resistor generates some noise, so reaching a noise free state is impossible given the current state of physics in the universe. To perform at all, electronic equipment such as Loran and SatNav units must be very sensitive to electromagnetic waves. Any spurious generation of noise can and does affect the performance of such equipment.

Noise comes in two basic varieties ... *radiated* and *conducted*. Radiated noise consists of magnetic waves which are dispersed in all directions from a source. For instance, a radio antenna radiates electromagnetic waves into the atmosphere. Conducted interference is that noise which results when unwanted current is conducted in series

with the power supply leads of sensitive electronics. Any current conducted through wires ends up with some amount of radiated energy,
so it is imperative to stop conducted noise before long wire runs. If
not, a long antenna results.

In all cases, it is preferable to stop noise at the source, rather
than attempt to limit its ill effects later. Radiated noise can only
be contained within the enclosure which houses the source. This is
best done with a complete metal enclosure. Even small holes in an
enclosure can allow radiated noise to escape. Small cracks between
adjoining metal walls can function as slot antennas and emit considerable energy. Battery chargers which use phase controlled SCR
circuitry can be troublesome radiated noise sources. Filtering the output of a charger that is radiating noise will not reduce the noise, since
filters only treat conducted noise.

By far, the biggest noise source aboard a boat or RV is the running
alternator. While its radiated noise may not be in the frequency band
to cause trouble, the alternator conducts a great amount of noise into
the electrical system, which in turn radiates noise as a consequence.
An alternator filter can remove most of the conducted noise. As noted
earlier in the book, a noise filter used with the alternator should be
one designed with the performance alternator in mind, since a cheap
automotive filter may fail under heavy alternator operation.

A simple capacitor across the alternator can reduce noise in many
cases. For best high frequency effectiveness, a polypropolene, ceramic
or mica capacitor is required. It may be connected in parallel with
an aluminum electrolytic or tantalum capacitor. For ultimate performance, a complex filter should be used. It should be connected as
close to the alternator output as possible.

A complex filter consists of an inductor in series with the alternator output, with one or more capacitors across the inductor output.
Such a filter is shown in Figure 15.10. The inductor, L1 impedes
a change in current flow. Capacitors C1, and C2 impede a change
in voltage. When used together, the inductor and capacitor reduce
noise by opposing any rapid changes in either voltage or current. As
shown, two capacitors are used. Capacitor C1 is polarized, such as
an electrolytic or tantalum. It is a large value capacitor, and filters
lower frequencies than does C2. Capacitor C2 is a small ceramic or

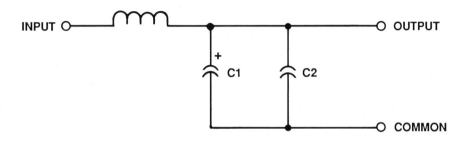

Figure 15.10: Schematic for a Complex Filter

mica unit which is most effective at higher frequencies.

Note that all the current must flow through the inductor, L1. Obviously for a high output alternator, a large inductor is required. At some *DC* current, any inductor saturates its magnetic core material and begins to act as an ordinary wire. Thus the inductor and core material must be selected for the application.

Where it is impossible to filter all of the electrical noise at the source, then a filter can be used preceding the electronics. An example of this was shown earlier in the discussion of the Charge/Access Diode. Such an electronic filter will also consist of an inductor with one or more capacitors. The inductor will be of a lesser rating than that required for the alternator.

15.7   Power and Grounding

Back in the 1960's we had the occasion to design a DC amplifier that was required to accurately amplify 1 uV (one millionth of a Volt) to a 1 Volt signal that could be shown on an analog meter. By the time the design became a production reality, we had learned a lot about electrical effects of temperature on dissimiliar metals, as well as the errors which result when minute currents flow in series with measurement leads. A wire is *not* a short circuit when low level DC measurements are considered. To reduce the effects of resistance in wires, it is necessary to separate current carrying circuits from sensing circuits. The ground for each of these circuits must be run separately to the power supply return point. Only by adhering to these rules can you expect sensitive electronic equipment to perform properly.

The order of connections does make a difference. Much of the problems that people have with electronics is the result of poor grounding practices. Indeed it is hard to imagine that a 3/8 inch brass distribution terminal bolt and the lugs attached to it are a resistive divider network, but it is a fact.

In Figure 15.11, a negative distribution terminal is shown. A 3/8 inch brass post should be used. Connected to the terminal post are various return wires from equipment. Note that the battery negative wire is placed on the terminal such that it separates the charge sources from the loads. The engine starter return is an exception to this rule. In both cases, however, the heavy current leads are placed closest to the battery return. Toward the outside, the smaller current carrying leads are placed.

At first glance this appears to be counter to logic. Why, for instance, aren't the most sensitive leads placed closest to the battery? As shown, the sensitive circuits are connected close together and no high current flows between them, or through them. They share a common reference. Indeed the common reference may fluctuate slightly from the actual battery lead as high current fluctuates beneath the low current wires, but all the sensitive equipment floats at a common potential. Noise thus generated is called *common mode* noise, and generally speaking, equipment is less sensitive to common mode noise than *normal* mode noise. If the load bearing leads were reversed, then

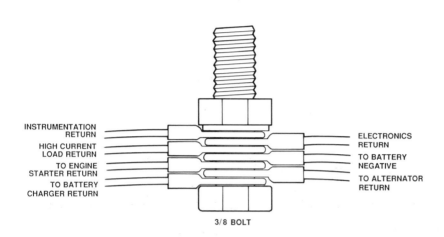

Figure 15.11: Negative Distribution Connections

the high current loads would be connected on the outside. Beneath them would be placed the returns from the electronics. Each of the sensitive return lugs would form a resistive divider, and as the heavy current loads fluctuated, each sensitive lead would take on a different value. Instead of sharing a common reference point, the sensitive leads would be driven at different potentials.

Suppose an antenna return was physically separated from the ground return of a Loran unit in such a way that the return current from an inverter flowed first through the antenna return and then the Loran return. The result would be a voltage potential between the antenna and Loran which would look exactly like a received signal ...noise.

As illustrated, the most sensitive return leads should be placed on the outside where they share a common reference.

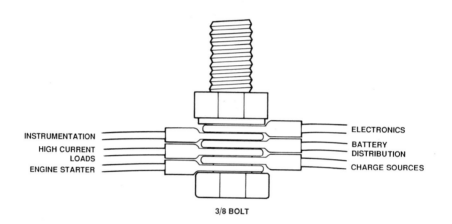

Figure 15.12: Positive Distribution Connections

While power leads are not generally as critical as ground return connections, it is still a good practice to connect power leads in the proper order. Power connections should be made in the opposite order than ground leads, that is, the most critical electronics should derive power closest to the battery. Figure 15.12 shows the way to intersperse the battery distribution lead with the connections to other equipment. Note that heavy loads are connected on one side of the positive distribution lug, while lighter loads are place on the other side. Connected in this manner, sensitive loads are not in series with high current loads.

Chapter 16

Symbols and Circuits

16.1 General Information

This chapter is devoted to symbols, circuits, mathematics and history.

At first, symbols always appear mysterious. Only after you have used the symbols for a while will they begin to take on meaning. We don't recommend that you learn these symbols ...just use this chapter for a reference when needed.

In this chapter we explain the simple mathematics which is sufficient to deal with simple *DC* circuits.

We also take some time to recite a little history of science and technology. Too often we hurry about our own lives without thinking about the lessons of the past. The puzzles which we see assembled today were once scattered pieces, and except for the hard work and brillance of those who lived before us, we might still be cave dwellers, dreaming about a balanced energy system.

16.2 Volts, Amps and Ohms

Figure 16.1 shows the symbol for a battery which is the short and long parallel lines. By convention, the longer of the lines is the positive terminal. Note that to measure the voltage of a battery, the voltmeter is placed across the battery terminals. The electromotive force *EMF* of a battery is measured in units of Volts, which is named after the

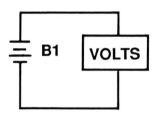

Figure 16.1: Battery with a Voltmeter

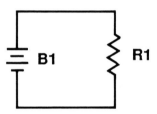

Figure 16.2: Battery with a Resistor

Italian physicist Alessandro Volta. He constructed batteries in 180(
which could conduct large currents.

Figure 16.2 shows the symbol for a resistor, which is the shor
diagonal lines. A resistor is named because it resists the flow of current. Resistance is measured in units of Ohms. The written symbo
for Ohms is the Greek letter Ω. Resistors are *loads* on a battery, s
anytime one wishes to show a load in general, the resistor is drawn
The unit of resistance is named after George Ohm, a German scien
tist. His discoveries in 1827 were held up to so much critcism tha
he was forced to resign his high school teaching position. He lived i
poverty until the late 1840's when his work was finally recognized.

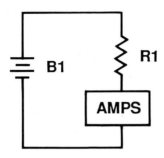

Figure 16.3: Battery with Ammeter

Figure 16.4: Symbol for a Fuse

Figure 16.3 shows how an ammeter (Amp meter) is used to mea-
ure current. The resistor is necessary, because a perfect ammeter is
:quivalent to a short circuit. Even an imperfect ammeter is close to
►eing a short. An ammeter can therefore never be connected directly
ιcross a battery. Note that before current can flow, there must be a
omplete circuit made across both terminals of a battery.

The unit of current is an Ampere. Like so many others, we have
ome to call it an Amp, but with no disrespect for André Ampère, the
'rench mathematician and physicist who first made the distinction
►etween Volts and Amperes. Ampère had mastered higher mathe-
natics by the age of 12, and in 1823 advanced the first theories of
:lectromagnetism.

Figure 16.8 shows the symbol for a fuse. This element is a much
ιeeded device, even when you don't experiment with circuits. We
lon't know the inventor of the fuse but reckon that the inventor was
ι practical person.

16.3   Ohm's Law

With the introduction of Volts, Amps, and Ohms, you are prepared
to digest the fundamental relationship of electronics. That of course
is Ohm's law. In words, Ohm's law states that the current flow is directly proportional to the applied voltage, and inversely proportional
to the resistance. In math symbology, Ohm's law is:

$$I = E/R$$

The I stands for the amount of current. We suppose that I is
used for historical reasons because current flow could *induce* motion
in magnetic compasses. In the equation above, the letter E represents
the number of Volts applied to the circuit. The letter E is short for
EMF. The letter R represents the resistance in the circuit.

Ohm's law can be rearranged by algebraic rules into two other
forms shown below.

$$E = IR$$
$$R = E/I$$

Note that two math adjacent symbols, such as I and R are to
be multiplied. Don't be confused by the absence of a multiplication
operator such as X used in grade school. Adjacent symbols which are
separated by parenthesis are also to be multiplied. Ohm's law could
be stated as $E = (I)(R)$.

Applying Ohm's law is straightforward. For instance, a battery of
12.5 Volts connected to a 5 ohm resistor will cause a current of 2.5
Amps to flow. $(I = E/R) = 12.5/5 = 2.5$)

Power consumed by a resistor (load) is expressed as:

$$P = EI$$

By substituting for E in the above equation, where $E = IR$, you
can also see that:

$$P = I^2 R$$

Figure 16.5: Capacitor

From our prior example for Ohms law, we had 12.5 Volts and 2.5 Amps so power being dissapted by R is $(12.5)(2.5) = 31.25$ Watts. If the circuit operates for 2 hours, we will have used 62.5 Watt-hours of energy. Because we had 2.5 amps of curent flowing for 2 hours, we will have used 5 Amps-hours from our battery.

The unit of power, Watts, is named after the Scottish inventor James Watt. Watt is usually credited with the *invention* of the steam engine, but actually he was the first to underestand and apply the principles of latent heat to an *existing* steam engine, making it so efficient that the engine could compete against horses used to lift water from mine shafts. This took place about 1769. He invented the term *horse power* and defined it as 550 foot-pounds per second, which was about the amount of work that the average horse could do. With this comparison, he could rate the amount of work that his steam engine could perform. Though Watt had nothing to do with electricity, electrical power is rated in Watts. There are 746 Watts in 1 HP.

16.4 Capacitors

Figure 16.5 shows the symbol for a capacitor. Capacitors store energy somewhat like a battery. Unlike the battery, however, there is no *DC* current through a capacitor.

A capacitor has a property called *capacitive reactance* which is inversely proportional to the applied frequency. At *DC* an ideal capacitor represents an open circuit. If *AC* voltage is applied to a capac-

itor, then the capacitor represents a load on the circuit. Capacitive
reactance is given by the equation:

$$X_c = 1/2\Pi F C$$

Units of F are Hz, and units of C (capacitance) are Farads. The
reactance of the capacitor is the term X_c. You can think of cpacitance
reactance much like resistance, except that voltage and current are not
in phase as they are through a resistor. In a pure capacitor, current
leads voltage by 90°.

The unit, Farad, is derived from Michael Faraday, one of the all
time great scientists. In 1805, fourteen year old Faraday was apprenticed to a bookbinder. Apparently he read more books than he bound.
In 1821, he was the first to convert current flow and magnetism to
continual mechanical motion, the DC motor. During the same period
of time Faraday invented the transformer, and demonstrated electrical induction. In 1823 as a chemist, Faraday discovered benzene, a
key compound that led to an explanation of molecular structure. In
1824, he discovered the principles of absorption refrigeration. In 1831
Faraday invented the electrical generator, which may be the root of
modern humanity. In 1832 he did the analysis of electro-chemical
reactions which take place in a battery. Later he made discoveries
of the interactions of light and electromagnetism. In the 1850's the
British government asked Faraday about the possiblility of making a
poison gas to use in the war against Russia. In a noble example of
concern for humanity, Faraday admitted the possibility of poison gas
but unequivocally refused to take part in its manufacture.

16.5   Inductors

Figure 16.5 shows the symbol for an inductor. The inductor is the
mathematical mirror or a capacitor.

At DC the inductor represents a *short* circuit rather than an open
circuit. At high frequencies, where the reactance of a capacitor is low,
the reactance of an inductor is high. Inductors also store energy, but
whereas a capacitor opposes changes in the applied voltage, the inductor opposes changes in the current flow. Used together, they oppose

Figure 16.6: Induct

changes to either voltage or current and thus *filter* noise. Inductive reactance (X_l) is given as:

$$X_l = 2\Pi FC$$

As above, units of F are Hz, but units of L (inductance) are Henries. This unit of measure is named after the American inventor, Joseph Henry. Henry was apprenticed to a watch maker at the age of thirteen, in 1810. He put himself through school and taught mathematics in 1826 at Albany, New York. A few years later he was building electromagnets, using his wife's silk stockings as insulation for the wire. In 1834, he discovered self induction in an inductor. In 1835, he invented the relay, and consequently the telegraph. (Morse is generally given credit for the invention of the telegraph, but it is Henry who helped Morse, a man quite ignorant of science.) Later Henry made significant contributions to the art of building electric motors, from whence flows much of our abundant lifestyle.

16.6 Diodes

Figure 16.7 shows the symbol for a diode, which used to be called a rectifier. The rectifier conducts current in only one direction.

A German physicist, Karl Braun discovered in 1874 that some crystals would conduct electricity in one direction better than the other. These crystals were later the basis of crystal radios. The practical rectifier was first invented by an English engineer, John Fleming

Figure 16.7: A Diode or Rectifier

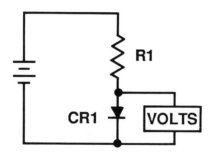

Figure 16.8: Forward Biased Diode

in 1904. The rectifier was made as a vacuum tube. Fleming called the
device a *valve*, because it would only conduct electricity in one direction. Modern diodes owe their existence to the research of William
Schockley, who's work of the 1940's culminated in the invention of
the transistor in 1948.

Figure 16.8 shows a circuit with a *forward biased* diode. The diode
conducts current in the circuit as shown. The resistor limits the current through the diode. The voltmeter across the diode will read 0.2–
1.5 Volts depending on the type of diode, and the amount of current
flow.

Figure 16.9 shows a circuit with a *reverse biased* diode. The diode
will not conduct current in the circuit as shown, unless the voltage is
higher than the rating of the diode. The voltmeter across the diode
will read the full voltage of the battery.

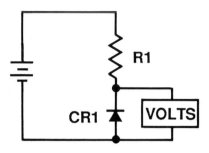

Figure 16.9: Reverse Biased Diode

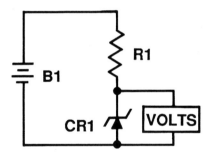

Figure 16.10: Zener Diode

Figure 16.8 shows a circuit with a *zener* diode. Note the extra lines in the symbol, as opposed to the normal diode. The diode conducts current in the circuit as shown as long as the battery voltage is higher than the *breakdown* voltage of the zener. The zener diode is designed to conduct in the reverse direction at a precise voltage, thus the zener is useful to derive a constant voltage. The resistor sets the current through the diode. The voltmeter across the diode will read the breakdown voltage of the zener.

The snubber sold by Ample Power Company[1] to protect the electrical system whenever the alternator is suddenly disconnected, is a

---

[1]Catalog # 1030.

special purpose zener diode. It is not designed for a constant voltage as much as for large transient current capacity. It's reverse breakdown voltage is about 23 Volts. The diode can conduct as much as 110 Amps at 23 Volts for a short period of time. Snubbers can be used in parallel for more current rating. Transient absorbers, such as the snubber, are rated in Joules of energy that they can dissipate. A Joule, you may recall, is equivalent to 1 Watt of power for 1 second, and is the International Standard Unit of energy measurement.

The Joule is named in honor of James Joule. Unlike many of our early scientists, Joule was born of wealthy parents, and could afford a life devoted to research. In 1840, he had determined the relationship between electrical current and the heat that it generated. On his honeymoon he took time out to measure the temperature of water at the top and bottom of a waterfall, believing correctly that the mechanical energy of the falling water would be translated to heat. This equivalence of mechanical energy and heat is the basis of the *first law of thermodynamics* which states that energy can neither be created nor destroyed, but can be changed from one form to another.

16.7   Transformers

Figure 16.11 shows the symbol for a transformer. Transformers are constructed by winding wire around a magnetic core. Alternating current in one winding (called the *primary*) causes alternating current to flow in the *secondary*. In the transformer, if we have 100 turns of wire around the core of the primary and 25 turns of wire around the core of the secondary, then, if the primary voltage is 115 Volts, the *AC* secondary voltage will be $(115)(100/25) = 28.75$ Volts. This would be a *stepdown* transformer as opposed to a *stepup* transformer. Note that primary and secondary circuits are isolated ... no direct connection for *DC* current. A center tapped transformer has $1/2$ the secondary voltage from the center tap to either side. Transformers are usually about 95% efficient in power transfer.

As noted earlier, the transformer was first invented by Faraday. It was Nikola Tesla, a Croatian engineer working in the U.S. that made the transformer a practical device capable of raising a low voltage to a

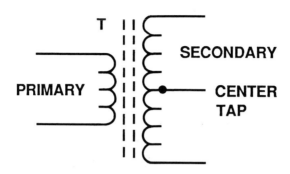

Figure 16.11: Transformer

very high voltage that could be transmitted long distances. Tesla and George Westinghouse had to battle the prestige of Thomas Edison to win the battle of AC power over DC power for the public utilities in the U.S. Edison fought AC power by first convincing the state of New York to use AC for its electric chair. Then he used the deadly nature of AC as an argument for DC.

You can see for yourself by looking at Ohm's law and the equation for power that for the same wiring resistance, higher voltages and smaller currents are more efficient than low voltages and high current. The transformer only steps up or down AC, not DC so Edison eventually lost to efficiency.

16.8 Wires on Schematics

Figure 16.12 shows two wires which cross on a schematic, which are *not* connected together.

Figure 16.13 shows two wires which cross on a schematic, which *are* connected together.

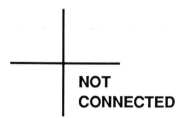

Figure 16.12: Un-connected Crossing Wires

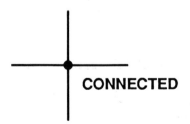

Figure 16.13: Connected Crossing Wires

Appendix A

Using the Wire Table:

You have a big anchor windlass, and the motor is rated at 1500 Watts. Since Power (P) = Volts (E) times Amps (I), find that 1500/12 Volts = 125 Amps. Since resistance = E/I (from Ohm's law), the motor resistance is 12/125 = 0.096 ohms. We want to get maximum power from the windlass, so lets start our design process by asserting a goal ... limit the wiring resistance to 1% of the motor resistance. With this goal in mind, wire resistance must not exceed 0.00096 Ω.

Wire length from the batteries to the windlass, *and* return is 50 feet. Dividing 0.00096 by 50 yields 0.0000192 Ω per foot allowed. We see that AWG #00 is about 4 times this value. If we use AWG #00 wire then 50 feet of it will be 0.00405 Ω. Dividing this by the motor resistance shows that the wire will lose about 4% of the available energy. We accept this compromise because good design practices limit wire loses to about 5%. We would have to use 4 wires to attain our 1% goal, and as it is, the weight of the wire is 20.5 pounds.

Big wire tip:

Often, you can find AWG #0, and AWG #00 wire at industrial surplus outlets. They sell wire that was used for welding, but has become unserviceable due to small insulation nicks, or bad terminals.

The wire is made up of many more finer strands than that normally sold for marine use, and will have lower resistance as a consequence. Select the best pieces, and plan on cutting at the insulation breaks. If you choose to repair the insulation, do so with a dab of silicone sealant, followed immediately with a double wrap of black tape. Then add an *over coat* of heat shrink tubing which can be obtained at most electronic suppliers.

American Wire Guage	Ohms Per foot	Weight Per Foot	Current MaxAmps
18	0.00651	0.00492	5
16	0.00409	0.00782	10
14	0.00258	0.0124	20
12	0.00162	0.0198	25
10	0.00102	0.0314	30
8	0.000641	0.0500	50
6	0.000403	0.0795	70
4	0.000253	0.126	90
2	0.000162	0.205	125
0	0.000102	0.326	200
00	0.0000811	0.411	225

Table A.1: Copper Wire Table – 77°F

Note: Maximum Amps is that current which will overheat the insulation if continuously loaded.

Multiply	By	To Obtain
Btu	778.2	foot-pounds
Btu	1055	Joules
Btu per minute	17.58	Watts
Btu per ft^2 per minute	0.1221	Watts per inch^2
feet^3	1728	inch^3
feet^3	7.481	gallons
feet^3	28.32	liters
inch^3	0.0005787	feet^3
inch^3	0.004329	gallons
inch^3	0.01732	quarts
fathoms	6	feet
feet	0.3048	meters
foot pounds	0.001285	Btu
foot pounds	1.356	Joules
gallons	0.1337	feet^3
gallons	231	inch^3
gallons	3.785	liters
horse power	42.4	Btu per minute
horse power	745.7	Watts
inches	2.54	centimeters
inches of mercury	0.03342	atmospheres
inches of mercury	0.4912	pounds per inch^2
Joules	0.0009480	Btu
Joules	0.0002389	kilogram-calories

Table A.2: Conversion Factors, Part 1

Multiply	By	To Obtain
kilogram-calories	3.968	Btu
kilogram-calories	4186	Joules
kilometers	3281	feet
kilometers	0.6214	miles
kilowatt-hours	3413	Btu
liters	0.03531	feet^3
megohms	1,000,000	ohms
meters	3.281	feet
micro-farads	0.000001	farads
miles	5280	feet
miles, nautical	6080	feet
mils	0.001	inches
ounces, fluid	1.805	inch^3
ounces, fluid	0.02957	liters
pounds	16	ounces
pounds of water	0.01602	feet^3
pounds of water	27.68	inch^3
pounds of water	0.1198	gallons
feet^2	144	inch^2
miles^2	640	acres
yards^2	9	feet^2
tons	2000	pounds
Watts	0.05688	Btu per minute
Watt-hours	3.413	Btu

Table A.3: Conversion Factors, Part 2

Material	Pounds per foot^3
air $(0^\circ F)$	0.0809
alcohol	49–57
aluminum	168
brass	510–542
bronze	545–555
cement, set	168–187
charcoal	17–35
coal	44–88
copper	554
cork	15.6
glass	150–175
gold	1203
ice	55–57
iron	440–490
lead	710
oil, petroleum	55
platinum	1336
salt	129–131
silver	660
soapstone	162–175
steel	474–494
tin	455
titanium	187–190
water, fresh	62.4
water, sea	64.0–64.3
wood, cedar	37–38
wood,lig vitae	78–83
wood, mahagony	32–53
wood, spruce	25–32
zinc	448

Table A.4: Weights of Common Materials

Material	Specific Heat
alcohol	0.615
brass	0.0917
bronze	0.0860
cement	0.271
coal	0.201
cork	0.485
cotton	0.362
diesel fuel	0.504
gasoline	0.500
ice	0.505
iron	0.119
lead	0.0297
marble	0.206
steel	0.118
water	1.000
wood	0.420
wool	0.393
zinc	0.0950

Table A.5: Specific Heat of Common Material

Material	Specific Heat Btu per ft^2 per inch per $°F$ per hour
air $(0°C)$	0.165
brass	592.21
brick	4.355
concrete	6.39
copper	2664.95
cotton	0.319
diesel	1.13
earth, typical	11.6
feathers	0.464
flannel	0.10
gold	2032.1
ice	11.322
iron, cast	316.4
lead	240.95
marble	24.4
paper	0.90
polyurethane	0.17
sand, dry	2.5
water 30°C	4.64
wool	0.406
zinc	769.3

Table A.6: Thermal Conductivity of Common Material

Bibliography

Issac Asimov, "Asimov's Biographical Encyclopedia of Science and Technology: the Lives and Achievements of 1510 Great Scientists from Ancient Times to the Present Chronologically Arranged", Second Revised Edition, Doubleday and Company, Inc.: Garden City, New York, 1982.

M. Barak, "Electrochemical Power Sources: Primary and Secondary Batteries", First Edition, Institute of Electrical Engineers (IEE) Series 1, The Institute of Electrical Engineers: London and New York, Peter Peregrinus Ltd., Stevenage, UK, and New York, 1980.

A. E. Fitzgerald, Charles Kingsley, Jr., Alexander Kusko, "Electric Machinery: The Processes, Devices, and Systems of ElectroMechanical Energy Conservation", Third Edition, McGraw-Hill, Inc.: New York, 1971.

G. Smith, "Storage Batteries: Including Operation, Charging, Maintenance and Repair", Second Edition, Pitman Publishing: Great Britain, 1971.

William M. Flanagan, "Handbook of Transformer Applications", McGraw-Hill, Inc.: New York, 1986.

E. M. Pugh, E. W. Pugh, "Principles of Electricity and Magnetism", Second Printing, Addison-Wesley Publishing Company, Inc.: Reading, Massachusetts, 1962.

John A. Aseltine, "Transform Method in Linear System Analysis", McGraw-Hill Book Company, Inc.: New York, 1958.

Phillip Cutler, "Electronic Circuit Analysis; Volume 1 Passive Networks", McGraw-Hill Book Company, Inc.: New York, 1960.

Curtis Instruments, Inc., "Battery Book One; Lead Acid Traction Batteries", Curtis Instruments, Inc.: Mt. Kisco, New York, 1981.

Conrad Miller and Elbert S. Maloney, "Your Boat's Electrical System; Including A Manual of Marine Electrical Work", Second Edition, Hearst Books: New York, 1981.

Philip Scorn, E. R. Ambrose, Theodore Baumeister, "Heat Pumps", Second Printing, John Wiley & Sons, Inc.: New York, 1947.

Edwin P. Anderson, Revised by Rex Miller, "Refrigeration: Home and Commercial", Third Edition, First Printing, The Bobbs-Merril Company, Inc.: Indianapolis, Indiana/New York, 1984.

Guy R. King, "Basic Refrigeration; Principles, Practice, Operation", Nickerson & Collins Company: Chicago, Illinois, 1951.

Electro-Craft Corporation, "DC Motors, Speed Controls, Servo Systems," Fifth Edition, Electro-Craft Corporation: Minnesota, 1980.

Andrew D. Althouse, Carl H. Turnquist, Alfred F. Bracciano, "Modern Refrigeration and Air Conditioning", Fourth Edition, The Goodheart-Willcox Company, Inc.: South Holland, Illinois, 1982.

Air Conditioning and Refrigeration Institute, "Refrigeration and Air-Conditioning", Prentice-Hall, Inc.: New Jersey, 1979.

General Electric, "The Sealed Lead Battery Handbook", Publication BBD-OEM-237, General Electric Company: Gainesville, Florida, 1979.

American Boat and Yacht Council, "Standards and Recommended Practices for Small Craft": Amityville, New York, 1987.

Motorola, "Linear/Switchmode Voltage Regulator Handbook: Theory and Practice", Second U.S. Edition, First Printing, Motorola, Inc.: 1982.

Nigel Warren, "Metal Corrosion in Boats", International Marine Publishing Company: Camden, Maine, 1980.

David Linden, "Handbook of Batteries & Fuel Cells", McGraw-Hill Book Company: New York, 1984.

Edgar J. Beyn, "The 12 Volt Doctor's Alternator Book", Spa Creek, Inc.: Annapolis, Maryland, 1986.

George Chryssis, "High-Frequency Switching Power Supplies: Theory and Design", McGraw-Hill Book Company: New York, 1984.

Underwriters Laboratories, "UL 1112: Marine Electric Motors and Generators (Cranking, Outdrive Tilt, Trim Tab, Generators, Alternators)": Northbrook, Illinois, 1983.

Underwriters Laboratories, "UL 1236: Battery Chargers": Northbrook, Illinois, 1986.

Nigel Calder, "Refrigeration for Pleasure Boats", Nada Enterprises, Inc.: Hammond, Louisiana, 1986.

"Dole Complete Catalog", Dole Refrigerating Company: Lewisburg, Tennessee; Dole Refrigerating Products Limited, Oakville, Ontario, Canada, 1987.

Alco Controls, "Catalog 23: Product Handbook of Specifications Applications and Selection of Refrigerant Flow Controls", Alco Controls Division: St. Louis, Missouri, 1984.

Sporlan, "Condensed Catalog 201", Sporland Valve Company: St. Louis, Missouri, 1986.

Sporlan, "Bulletin 40-10: Catch-All Liquid Line and Suction Line Filter-Driers Plus AK-1 and AK-3 Acid Test Kits", Sporland Valve Company: St. Louis, Missouri, 1984.

Michael W. Dempsey, "Illustrated Fact Book of Science", First Edition, Arco Publishing, Inc.: New York, 1983.

Jane D. Flatt, Publisher, "The World Almanac and Book of Facts 1982", annual publication by Newspaper Enterprises Association, Inc.: New York, 1981.

Jane D. Flatt, Publisher, "The World Almanac and Book of Facts 1985", annual publication by Newspaper Enterprises Association, Inc.: New York, 1984.

Maurice V. Joyce and Kenneth K. Clarke, "Transistor Circuit Analysis", First Edition, Addison-Wesley Series in the Engineering Sciences, Addison-Wesley Publishing Company, Inc.: Reading, Massachusetts, 1961.

Arthur W. Judge, "Modern Electrical Equipment for Automobiles", Volume 6, Motor Manuals: A Series for all Motor Owners and Users, First Edition, Taylor Garnett Evans and Company, Ltd.: Great Britain, 1962.

Ray F. Kuns, "Automotive Essentials", Revised Edition, Bruce Publishing Company: Milwaukee, Wisconsin, 1962.

Jack Park, "The Wind Power Book", First Edition, Cheshire Books: Palo Alto, California, 1981.

Hans Thirring, "Energy for Man: From Windmills to Nuclear Power", First Edition, Harper Colophon: New York, 1976.

R. L. Watts, S. A. Smith, J. A. Dirks, R. P. Mazzucchi, "Photovoltaic Product Directory and Buyers Guide: Home, Farm, Industrial, and Institutional Systems, Components, and Appliances", Second Edition, Van Nostrand Reinhold Company: New York, 1984.

Ralph G. Hudson, S.B., "The Engineers' Manual", Second Edition, John Wiley & Sons, Inc.: New York, 1939.

National Fire Protection Association, (Safety information and standards), Boston Massachusetts, 1987.

Index

absorbed electrolyte 49
absorption 27,31,41,44
   natural 31,52
absorp. rate/refri. 233,238
AC, grounding 259
   system 255
   waveform 129
accuracy 277,278
acid diffusion 23
acid, loss 28
air, circulation 145
   leaks 147
   motionless 155
Albany, New York 307
alkaline 189
alternator 43,61,78,253
   cold 67,83
   disconnected 71
   efficiency 81
   failure 68
   hot rating 67
   HP 81
   kkk 68
   large frame 66
   lazy 79
   load sharing 79
   main 252
   N Type 70
   noise suppression 74

output 65,66,74
P Type 70
PN type 70
parallel 78,79
re-assembly 71
reliability 67
small frame 83
spares 88
testing 86
wimpy 50
aluminum 89,174
Am. Boat & Yacht Cnsl. 268
ammeter 132,303
ammonia 153,154,196
   latent heat 154
   pressure of liquid 154
   SG 154
Ampere-hours 15
Ample Power Co 52,68,78,84
   110,111,251
amplifiers, current 278
   stabilized 279
Amps, cold cranking 19
Ampère, André 303
anode 11
antennas, slot 294
anti-freeze 190
antimony 30,40,49
appliance, ungrounded 264

arcing 86
arsine 40
bacteria, growth 145
Baja California 157,239
baking soda 41
batteries, equal banks 43
    Ni-Cad 47,57
    sealed 47,250
    size 184
    charger 91
    life 33
    symbol 301
battery, 18
    automotive 12
    banks 250
    cold 14
    continuity 54
    cost 56
    deep cycle 12
    new 18,19,28,33,37,113
    old 33,37,113
    rechargeable 11
    rested 110
    sealed 12
    series, weaker 37
    sizing 250
    wear out 58
bearings, 67,83
    failure 88
    installation 89
    pullers 89
    removal 89
belt, dressing 85
    loss 81
    tension 85
    wrap 85
benzene 306

blank periods 33
bond 262,269
bore and stroke 242
box, transfer 147
braided shield wire 74
Braun, Karl 307
breaker, rating 130
    spare 130
brightness control 275
brine 186,188,191,207
British Thermal Units 149
brush, alternator 61
brushes, failure 86
Btu 149
bulb, sensing 179,223
calcium chloride 189,209
capacitive reactance 305
capacitor, ceramic 75
    electrolytic 270,288
    metallized mylar 74
    mylar film 74
    polypropolene 74
    symbol 305
capacity 15,18,28,37,52
    50% 42
    box 239
    brine 186
    compressor 242
    delivered 28
    excess compressor 187
    needed 19
    remaining 21,25,43
    reserve 19
    test 19,20,23,26
    to-weight ratio 18,19
capillary tube 177
car, electric 12

arbon brushes 61
ases, phenolic 18
    polypropylene 18,49
athode 11
ell failure 38,39
ell reversal 57
entrifugal pump 244
harge 50,28
    absorption 27,50,84
    acceptance 28
    bulk 27,28,29,34,84
    correct 26
    efficiency 29
    equalization 27
    excessive 39
    fast 28,101
    float 27,50,54,84
    full 28,31,37,38,40,44
    50,52,101
    slow 14,29,44
    state of 21,52
    switch 75
Charge/Access Diode 285
harger 96
    automotive 94
    choosing 101
    cycling 99
    dockside 102
    efficiency 95
    ferro-resonant 96
    129,134
    multi-cycle 100
    phase controlled 97
    specifications 92
    unregulated 95
harger/battery match 100
harging

absorption 28,44,69,84
    equalization 54
    fast 51
    float 84,111
    individual 67
    offshore 101
    performance 26,29
    proper 84
charter 72,84,85
Chinese 144
chromate 189
circuit breaker 130
Coast Guard 244
compression ratio 204
compressor 163,166
    184,218
    auto air 210,240
    capacity 239
    damage 248
    DC 187
    destruction 163,181
    hermetic 166
    HP 243
    intake 163
    load 185
    oil 169
    open 166
    rating 170
    rotation 204
    run time 240
    seals 168,248
    shut-off 238
    small 155
    timer 187
    valve 166
condenser 163,170,214
    air 164,232

cleaning 173,244
cooling 172
copper 170
efficiency 170,244
pressure 172
tube/tube 173,215,244
volume 172
water 164,232,240
condition, testing 20
conductivity, thermal 150
conductor, human 264
constant voltage 54
constant current 54
consumption, Amp-hours 19
daily 19
Continental Edison Co. 121
continuous cycle 184
control 68,246
controls, shut-off 247
conversion efficiency 107
conversion rate 276
converter 133
cooling fins 172
cooling, measurement 225
copper tubing 180
copper wire 67
corrosion 31,40,49
plate 30,42
cost, energy 20,229
coupling, flexible 81
crankshaft 166
cross-sectional area 212
crossover valves 242
cruising equipment 264
crystal radios 307
cupronickel 173,217,244
current

alternating 121
cathodic 269
charge limit 29
constant 37,66,100
excessive 33
float 54
flowing 122
impressed 266
inrush 138
leakage 264
lethal 125,266
limiting 66,253
measurement 251,278
minute 296
pulsed 32
reverse 286
sensor 75
sharing 38
starting 128
stray 264
test 20
transient 310
trickle 33
cycle life 15,51
cylinder 163
Dassler, Adolf 47
defrost 148
Delco 12,154
dendrites 41.
design approach 230
diffusion 13,15,21,27
28,29,50,51
digital readings 257
DIN 43539 51
diodes, 61,292
bridge 270
Charge/Access 110

drop 281
failure 68
forward biased 308
isolating 106
isolation 108,280
rating 76
reverse biased 308
Schottky 68,76,106,119
short circuit 67
silicon 68,76
trio 61,70
tunnel 68
zener 309
dirty batteries 41
discharge 12,18,28,32,51
  50% 19,20,23,40,43
  51,108,250
  deep 38,51,58
  excessive 39
  fast 14
  full 23
  intermittent 13,43
  rate 15
displays 273
  power 276
  readability 275
  segment 277
distilled water 40,48
Dole Refrigerating Co. 233
dual Schottky diode 285
E i Dupont & Co. 153
Edison, Thomas 11,12
  121,311
efficiency, absorption 199
  compressor 243
  SCR 98
  thermoelectric 196

transformer 310
volumetric 204,219
eggs, storing 207
Egyptians 144
electric chair 311
electrical noise 69
electrocution 264
electrode 11
electrolysis 73,95,102,129
  130,259,268
electrolyte 11,12,14,39,40
  41,48,49,50,54,253
  gelled 50
  immobilized 13
electromagnet rotation 66
electromagnetic waves 293
electromotive force 301
electrons 106,122
empirical 150
energy, latent 151
energy, refrigeration 164
engine
  block 73
  governor 83
  sizing 83
  small 81
  twin 79
enthalpy 202,218
entities 148
epoxy 292
equalization 14,18,27,34
  37,41,44
  independent 34
  monthly 41
  network 38
  problems 35
  resistors 35

time 35
error digit 278
Europe 271
eutectic 189
  brine 175
  freezer 233
  point 233,238,239,248
  reefer 236
  solution 232
evaporation 151,153
evaporator 163,174,199,208
  circulation 175
  flat plate 166
  starved 164
  surface area 175
expansion valve 163,177
  213,223,245,246
  valve blockage 181
  valve calculations 220
explosion 31
explosive, hydrogen 39
Fail Safe Diode 68,79
  281,252
failure, electrical 67,85
failure, pending cell 39
Faraday, Michael 11,144
  196,306
FCC 139
feedback 137
ferro-resonant effic. 97
field
  control 65
  current 66
  drivers 69
  winding 61,86
  wires, joining 79
filter, alternator 73,294

complex 288,294
simple 288
filter/drier 181
fins 186
Fleming, John 307
float 31,32,33
  current 33
  equipment 32
  voltage 54
floating, definition 32
flux 66,122
  magnetic 124
freezer load 235
freezer, storage 233
  top loading 235
Freon 153
frequency, adjust 129
  inverter 134
  operating 99
  sensitive to 129
frost 148
fuse symbol 303
gallons per minute 215
Galvani, Luigi 11
galvanic action 261
gas
  laws 151,153
  point 28,31,34,39,42,68
  tables 201
  transfer 50
gassing 28,29,40
GE–Intersil 276
gel, immobilizing 189
General Electric Co. 144
General Motors 12
generator 95,126
  auxiliary 129

electro-magnetic 122
portable 102,256
sizing 126
synchronous 122
lycol 190,232
rid 12,30,41,49,51
round fault isolator 133
rounding practices 296
AM 269
armonics 135
eadering 175
ealth hazards 40
eat
    exch./accum. 181,245
    charging 28
    compressor 169
    condenser 170
    definition 148
    latent 151,152,186,305
    measurement 149
    movement 164
    of plates 29
    quantity 149
    sensible 150
    transfer 148
    transfer, air 214
    transfer, water 215
eatsink 76,119,285,292
eatsink, calculations 293
enry, Joseph 307
ertz, Heinrich 122
igh voltage 71
oldover
    calculations 208
    cooling capacity 208
    plate 166,185,246
    plates 229

storage 186
horse power 305
HP, alternator 81
hum, audible 98
hydrogen 28,31,35,39,49,197
hydrometer 14,21,23
ice cream 164,210
ice harvest 144
ice melt 151
inches of mercury 225
indicator, polarity 130
inductance 69
inductive reactance 307
inductor 128,294,306
    saturation 295
instrumentation 25,38,44
    101,183,246,247,252
    refrigeration 166
insulation 155
    loss 157
    loss measurement 160
    rating 155
interference tests 139
interference, radio 139
intermittent cycle 185
Int'l System of Units 149
Intersil 276
inverter 126,133,138
    139,255
    rating 138
ion 262
iron 65,66
isolation transformer 271
isolator diodes 75,283
jargon 148
Joule, definition 149
Joule, James 310

k factor 157,236,150
Kelvinator 144
Kelvin, degree 149
kerosene, burners 197
Kettering, Charles 12,154
lamination, steel 66,93,98
LCD 275
lead 12,13,14,18,54
lead sulfate 13,41
lead/antimony alloy 49
LED 275
level, acid 40
life expectancy 32
life, battery 32
lightning 269
liq. line filter/drier 245
liquid metering 177
lithium bromide 197
load
    AC 126
    analysis 201
    electrical 248
    freezer use 236
    side 83
Loran 287,297
magnetic
    failure 88
    flux density 121
    intensity 66
    material 99
    poles 66
magneto-caloric 152
maintenance 37,40
manual regulator, using 84
material, active 30
measurement accuracy 276
measurement, cooling 225

mechanical energy 149
medicine 143
meter, Amp remote 74
meter, digital 19,132,276
Mexico 117,160,187,273
microwave oven 140
Midgley, Thomas 154
mist, acid 41
moisture sensing 248
Monitor/Regulator 251
Morse, Samuel 307
MOSFET 99,133
motor
    1 HP 128
    12-Volt 239
    AC 121,132
    burnout 169
    compressor 241
    cooling 169
    efficiency 243
    hermetic 132
    induction 138
    inverter 138
    knock 140
    permanent magnet 116
    synchronous 134,138
    universal 128
motor/compressor 241
multiplexer 278
N Type 61
Nat'l Fire Prot. Assoc. 268
negative distribution 296
negative plates 12
network, equalization 35
nickel cadmium 100
nobility series 261
noble 260

noise
    alternator 294
    charger 98
    common mode 296
    conducted 293
    containment 73
    electrical 69,100,293
    harmonic 135
    normal mode 296
    radiated 293
nylon bushings 73
Ohm's Law 304
Ohm, George 302
ohmmeter 70,86,88
Ohms 70
oil 182
    level 183
    dissolved 177
    insulator 177
    pump 169
    refrigerating 170
    trapped 182
open circuit voltage 32
oscillator, crystal 134
oscilliscope 129,256
output 65,66
overcharge 30,49
oxidation 30,44
oxygen 28,30,31,35,39,44,49
P Type 61
parallel alternators 78,80
parallel batteries 37,38
patent additives 41
Peltier 193
Perkins, Jacob 144
petroleum jelly 41
phase control 97

phase transition 151
photo-voltaic cells 105
photon 106
piston, oil grooved 166
plate
    cooling capacity 209
    negative 14,50
    positive 14,30,44,49
    51,54
    thick 29,43,49,73
    thin 28,49,73
    parallel 211
plating, galvanized 273
platinum catalyst caps 35
poison gas 306
polarity 86,130
polyurethane foam 156
positive distribution 299
positive plates 12
power
    calculations 242
    equation 304
    factor 126,128
    filters 73
    instantaneous 139
pressure
    absolute 224
    atmospheric 153
    battery 54
    compressor 168
    cut-out 217
    drop 222
    gauge 224
    liquid line 247
    rise 204
    sensing 187
    suction 247

transducer 225,247
Priestly, Joseph 154
propane 199
propeller, unbalance 118
propeller, water 116
propeller, wind 116
Puerto Escondido 117,164
Puerto Vallarta 273
pump
  bilge 287
  centrifugal 172,217
  oil 169
  water 216,240
pumpdown rate 186,210,213
pumpdown, freezer 236
pumpdown, rapid 186
pure DC 72,99
push tabs 85
R–12 154
  latent heat 154
  pressure of liquid 154
  SG 154
radar 248
radio freq. emissions 74,98
rate of discharge 15
rate, 20 hour 20,23
rate, charge 28
receiver 163,180
recomb. cap 31,35,39,56,101
recombination, oxygen 50
recovered energy 236
recovery phenomenom 13
rectifiers, silicon 61
redundancy 242
reefer load 238
refrigerant 153,175
  flooding 164

metering 163
velocity 211
refrigeration
  absorption 196
  absorption rate 232
  calculations 201
  DC 168
  design 201
  domestic 144
  effect 202
  efficiency 164
  electric 231
  habits 144
  hose 180
  load 204
  moisture 181
  oil 170
  organization 145
  power 252
  reference 164
  ton 244
refrigerant friction 213
regulation
  active 69
  automatic 100
  mode 66
  multi-step 66
  requirement 100
  series 118
  shunt 118
  stability 93
  types 93
  wind 117
regulator
  3-Step 78,85,111,253
  adjustable 112
  alternator 29

automatic 54,84
automotive 83
emergency 86
external 70
integral 86
internal 71
linear 111
manual 84
Multi-Source 84,253
multi-step 76,89
remote 71
sensing 72
series 111
shunt 111
solar panel 108
spare 89
special purpose 29
switched mode 111
with isolators 76
release valves, Bunsen 50
reliability, solar cell 107
resistance, battery 49
resistor symbol 302
resolution 277
reverse polarity 38
rings, piston 166
rotor 65,66
RPM, frequency 129
SAE 68
safety 39,129
AC 124
valves 54
saltwater, resistivity 259
saturation 96
Schauer Mfg. Corp. 94,254
Schockley, William 308
SCR 97

Sea of Cortez 232,248
sealed technology 48
seals, compressor 168
failure 187
leak 168
problems 169
Seattle, WA 264
Seebeck, Thomas 193
self-discharge 28,40,41,49
51,54,56,112
self-priming 217,245
sensing bulb 179,223
separators 12,29,40
felt mat 48
fiberglass 12,13,49
micro-porous 12,49
series connection 35,37,38
service fitting 225
shield 74
shock, deadly 264
shock, exposure 266
short circuits 12
shunt 19,111,132
current 74,278
magnetic 96
resistive 111
side load 83,88
sight glass 183,248
silicon
atoms 106
cell 107
amorphous 107
crystalline 107
silver chloride 196
sine wave 124
sine, trigonometric 124
slip ring 86

slow blow fuse 286
snubber 72,90,309
sodium chloride 189
solar
  cell 106
  panel 105,106,290
  panel, self reg. 110
  panels 114,254
  tracker 113
soluable, sulfate 41
spare alternator 89
specific gravity 14,21,23
  37,40,52
specific heat, def. 150
specification, proper 92
specify, definition 92
speed of light 13,122
spring pressure 224
square wave 135
SSB 269
stabine 40
Standard HDBK-217D 67
state of charge 258
state, change of 151
stator lamination 66,88
stator winding 65
steel laminations 66,93,98
stepped wave 136
storage mechanism 229
storage, out of service 41
strap, copper 269
stratification 190
stratify, electrolyte 54
suction filter 169
sulfate clumps 18
sulfate crystals 18,41,112
sulfation 19,32,48

sulphuric acid 12,13,14
  23,39
sunshine, energy of 105,107
superheat 224
swash plate 166
switch
  battery 76,108
  charge 75
  emergency 86
  ignition 86
  pressure 183
  selector 67,72,290
  transfer 129
switcher power supplies 99
symbols 301
synthesized sine wave 136
tachometer 129
tank, brine 238
temperature 14,15
  alternator 66
  Baja 232
  battery 54,84
  charger 93
  compensation 21,33,84
  compressor 247
  condenser 247
  definition 148
  difference 233
  drift 278
  eutectic 189
  evaporation 233
  food 145
  limit 29
  magnetic 67
  measurement 193
  probe 246
  sensing 163

terminal voltage 37
terminals, positive 74
Tesla, Nikola 121,310
thermal conductivity 150
    155,157,223
thermal resistance 285
thermocouple 193
thermoelectric 152,194
thermometer 23
thin film 107
thixotropic gel 12,13
three phase 61
ton, refrigeration 222
top loading 206
topology 32,33,93
transducer 183
transfer box 147
transfer switch 130
transformer 66,93,94
    99,121,133,135,310
    isolation 95,102,264
    saturation 96
    stepdown 310
    stepup 310
transient supp. device 72
transistor, invention 193
transistors, bipolar 139
trap, water 147
trickle charge 33
trickle current 56,113
tubing length 213
turns ratio 94
U.S. 271
UL 95,266
unbalance, alternator 66
Underwriters Lab., Inc. 95
Uninterr. Power Supplies 47

useful load 206
valve, expansion
    adjustment 220
    equalized 220
    frozen 247
    throttling 244
veg., moisture loss 145
velocity 201,212
ventilation 37,39
venting, battery 54
Volt-Amperes 126
Volta, Alessandro 11,302
voltage
    battery cell 23
    constant 31,33,66,100
    conversion 271
    dual sensing 282
    highest natural 37
    measurements 25
    open circuit 25
    panel maximum 108
    regulation 33
    Schottky 76
    sensing 282
    silcon diode 76
    solar cell 107,108
voltmeter 21,25,129,277
Volts, 24 126
volumetric efficiency 204
    219,242
vortex tubes 152
water
    boil, 970 Btu 152
    distilled 40,48
    electrolytic 259
    intake 245
    resistivity 259

sea 226
trap 147
velocity 215
Watt, James 305
Watts 106,126
wave, sine 140
wave, square 140,256
Westinghouse, Geo. 121,311
wind
    force 115
    generator 115,290
    power 115
    velocity 115
wind/tow hookup 118
winding
    alternator 61,65,86
    failure 88
    field 65
    primary 310
    secondary 310
    stationary 122
    stator 65
winter, 1890 144
wire
    black 125,130
    green 125,130,262
    ground 125,130
    not short 296
    safety 271
    sizing 133
    white 125,130
wires, crossing 311
wiring, AC 129
zinc 262

Ample Analyzer Diskette

- Runs on PC compatible Computers

- Menu Driven for Ease of Operation

- Computes Daily Amp-Hour Consumption

- Computes Daily Amp-Hour Charge Capacity

- Performs Daily Amp-Hour Balance

- Computes *Time to Charge* for Input Scenarios

- Writes or Reads Work Sheets to/from Disk

cut here

. . . .

return address

stamp

Rides Publishing Company
2442 NW Market Street #43B
Seattle, Washington 98107

The Ample Analyzer is a program that allows naming and describing system Loads (discharges), Sources (chargers) and Batteries. From the entered data, daily energy balance is computed, based on the time each Load or Source runs. Given a Source and a Battery the program computes the time required to recharge. Calculates charge time required in both conventional lead-acid batteries and the new immoblized sealed batteries. User input data may be saved on disk and also read from the disk, allowing several system scenarios to be tested for an optimum system design.

Balanced Energy Catalog from Ample Power Company with wiring diagrams, application notes, and product specifications including instrumentation, 3-Step Regulators, alternators, Fail Safe Diodes, sealed batteries, Charge/Access Diodes and refrigeration systems.

Mail to: RIDES Publishing Company, 2442 NW Market Street, #43B Seattle, Washington 98107

Order Form			
Item	Price	Quantity	Extended Price
Book - Living On 12 Volts With Ample Power	$25		
Balanced Energy Catalog	$4		
Ample Analyzer Diskette	$30		
Book, Diskette and Catalog. Save $15!!!	$44		
Tax (Washington Residents)	7.9%		
Shipping & Handling			$3.50
Total			

Name: _____

Address: _____

City: _____ State: _____ Zip: _____

Phone: _____